Marc-

Here's a copy of the book we discussed that was co-authored/chaired by Marianne MacDonald with CDM. Hopefully the information inside will help navigate some of the IT planning that DWU is considering in the future, particularly the chapters on guidelines for setting up a master plan.

Please don't hesitate to contact Marianne or me if we can answer questions or help out.

-Adam Evans

214. 912. 0596

EvansAD@cdm.com

INFORMATION TECHNOLOGY IN WATER AND WASTEWATER UTILITIES

Prepared by the **Information Technology in Water and Wastewater Utilities Task Force** of the **Water Environment Federation®**

Z. Cello Vitasovic, Ph.D., P.E., *Chair*
Marianne L. MacDonald, Ph.D., *Vice-Chair*

Daniel P. Baker, P.E.
Michael Barnett, Ph.D.
Hardat Barran, P.E., MBA
Burcin Becerik-Gerber, DDes
Rich Castillon
Peter Craan, P.E., CAP
Thomas DeLaura
Rienk de Vries, BCE, MBA
George R. Freiberg
Donald Gray
Marla K. Hartson, PMP
David Henry, P.E.
Charles P. McDowell
Jon H. Meyer
Gustaf Olsson, Ph.D.
Steven M. Ravel, P.E., BCEE
Melanie Rettie
Randal W. Samstag
Nedeljko Štefanic, Prof. dr.sc.
Michael Waddell
Corey Williams, P.E.

Under the Direction of the **Automation and Information Technology Subcommittee** of the **Technical Practice Committee**

2010

Water Environment Federation®
601 Wythe Street
Alexandria, VA 22314–1994 USA
http://www.wef.org

INFORMATION TECHNOLOGY IN WATER AND WASTEWATER UTILITIES

WEF Manual of Practice No. 33

Prepared by the Information Technology in Water and Wastewater Utilities Task Force of the Water Environment Federation®

WEF Press

Water Environment Federation Alexandria, Virginia

New York Chicago San Francisco Lisbon London Madrid
Mexico City Milan New Delhi San Juan Seoul
Singapore Sydney Toronto

The McGraw-Hill Companies

McGraw-Hill books are available at special quantity discounts to use as premiums and sales promotions, or for use in corporate training programs. To contact a representative, please e-mail us at bulksales@mcgraw-hill.com.

Information Technology in Water and Wastewater Utilities

1 2 3 4 5 6 7 8 9 0 WFR/WFR 1 6 5 4 3 2 1 0

ISBN 978-0-07-173705-0
MHID 0-07-173705-7

Printed and bound by Worldcolor/Fairfield.

This book is printed on acid-free paper.

IMPORTANT NOTICE

The material presented in this publication has been prepared in accordance with generally recognized engineering principles and practices and is for general information only. This information should not be used without first securing competent advice with respect to its suitability for any general or specific application.

The contents of this publication are not intended to be a standard of the Water Environment Federation® (WEF®) and are not intended for use as a reference in purchase specifications, contracts, regulations, statutes, or any other legal document.

No reference made in this publication to any specific method, product, process, or service constitutes or implies an endorsement, recommendation, or warranty thereof by WEF.

WEF makes no representation or warranty of any kind, whether expressed or implied, concerning the accuracy, product, or process discussed in this publication and assumes no liability.

Anyone using this information assumes all liability arising from such use, including but not limited to infringement of any patent or patents.

About WEF

Formed in 1928, the Water Environment Federation® (WEF®) is a not-for-profit technical and educational organization with members from varied disciplines who work toward WEF's vision to preserve and enhance the global water environment.

For information on membership, publications, and conferences, contact:

Water Environment Federation®
601 Wythe Street
Alexandria, VA 22314–1994 USA
(703) 684–2400
http://www.wef.org

Manuals of Practice of the Water Environment Federation®

The WEF Technical Practice Committee (formerly the Committee on Sewage and Industrial Wastes Practice of the Federation of Sewage and Industrial Wastes Associations) was created by the Federation Board of Control on October 11, 1941. The primary function of the Committee is to originate and produce, through appropriate subcommittees, special publications dealing with the technical aspects of the broad interests of the Federation. These publications are intended to provide background information through a review of technical practices and detailed procedures that research and experience have shown to be functional and practical.

Contents

Chapter 3 Information Technology Planning

Chapter 4 Developing an Information Technology Program for a Municipal Agency

Chapter 5 Information Technology Capital Project Management

Chapter 6 Information Technology Systems—Processes and Practices

Chapter 7 Information Technology Security

Chapter 8 Organizational Aspects of Information Technology

Chapter 9 Critical Success Factors and Key Future Challenges for Information Technology in Water and Wastewater Utility Projects

Chapter 10 Case Studies

List of Figures

List of Tables

Table **Page**

List of Tables

Preface

The purpose of this Manual of Practice is to provide an overview of information technology (IT) within water and wastewater utilities. IT plays an important role in the management of utilities, and IT systems are used in almost every aspect of a utility.

Efforts to implement IT solutions in utilities and execute IT projects have not gone without their share of challenges and difficulties. It would be impossible to prepare a document that would include all of the information necessary to eliminate the inherent risks that are part of applying IT to a utility. It is a vast technical area; even leading IT professionals and visionaries in this field often struggle with, and are bewildered by, this rapidly evolving technology. The authors of this manual focused on creating a useful and practical summary from an enormous body of information, with the goal of including only content that is most practical to a utility professional.

This Manual of Practice was produced under the direction of Z. Cello Vitasovic, Ph.D., P.E., *Chair*, and Marianne L. MacDonald, Ph.D., *Vice-Chair*. The principal authors of this Manual of Practice are as follows:

Chapter 1	Z. Cello Vitasovic, Ph.D., P.E.
Chapter 2	Marianne L. MacDonald, Ph.D.
	Burcin Becerik-Gerber, DDes
	Peter Craan, P.E., CAP
	Gustaf Olsson, Ph.D.
	Randal Samstag
Chapter 3	Marla K. Hartson, PMP
Chapter 4	David Henry, P.E.
Chapter 5	Hardat Barran, P.E., MBA
	Ashraf Azmi
Chapter 6	Michael Waddell
	Michael Barnett, Ph.D.
	Thomas DeLaura
	Z. Cello Vitasovic, Ph.D., P.E.
Chapter 7	Peter Craan, P.E., CAP

Chapter 8 Rienk de Vries, BCE, MBA
Chapter 9 Z. Cello Vitasovic, Ph.D., P.E.
Chapter 10 Daniel P. Baker, P.E
Michael Barnett, Ph.D.
Marla K. Hartson, PMP
David Henry, P.E.
Nedeljko Štefanic, Prof. dr.sc.
Z. Cello Vitasovic, Ph.D., P.E.

In addition to the WEF Task Force and Technical Practice Committee Control Group members, reviewers include Elizabeth Brandel, Alicia Crumpton, and Todd Jacob.

Authors' and reviewers' efforts were supported by the following organizations:

Barran Engineering, Inc., Toronto, Ontario, Canada
Beacon 2020, Inc., Oakville, Ontario, Canada
Carollo Engineers, Fountain Valley, California; Walnut Creek, California; Phoenix, Arizona; and Seattle, Washington
CDM, Cambridge, Massachusetts
City of Goodyear Wastewater Services Operations Superintendent Water Resources Dept., Goodyear, Arizona
East Bay Municipal Utility District, Oakland, California
Elsinore Valley Municipal Water District, Lake Elsinore, California
EMA, Inc., Longwood, Florida, and Saint Paul, Minnesota
Fakultet Strojarstva i Brodogradnje, Zagreb, Croatia
Hazen and Sawyer, P.C., Detroit, Michigan, and New York, New York
IEA, LTH, Lund, Sweden
Inflection Point Solutions, LLC, Overland Park, Kansas
Johnson County Wastewater, Olathe, Kansas
Knowledge Automation Partners, Inc., Wellesley, Massachusetts
Lee County Utilities, Fort Myers, Florida
Metropolitan Water District of Southern California, Los Angeles, California
Orange County Sanitation District, Fountain Valley, California
Reid Crowther Consulting, Bellevue, Washington
Richard P. Arber Associates, Lakewood, Colorado

2 Cats Consulting LLC, Phoenix, Arizona
Union Sanitary District, Union City, California
University of Southern California, Los Angeles, California
Westin Engineering of Michigan, Detroit, Michigan

INFORMATION TECHNOLOGY IN WATER AND WASTEWATER UTILITIES

Chapter 1

Introduction

1.0 OVERVIEW

During the past few decades, information technology (IT) has increasingly been incorporated into all aspects of utility operations, and, in the last decade in particular, technology has become a prominent part of everyone's lives. To respond to new business demands and technological trends, utilities continue to make significant investments in IT. As a result, IT plays a critical role in the management of most utilities today.

The purpose of this manual of practice (MOP) is to provide an overview of IT within water and wastewater utilities. One of the reasons that a similar MOP was not written earlier is that, in spite of the importance of IT, addressing the topic always seemed like a daunting task. Indeed, the IT field is vast and rapidly changing; new technologies and innovations, such as computer networks, continually emerge and change the way we live and work.

Even leading IT professionals and visionaries often struggle with, and are bewildered by, this rapidly changing field. In 1995, Microsoft's (Redmond, Washington) Bill Gates published a book called *The Road Ahead,* describing his vision of the future in IT. Out of 300 pages in the first edition of the book, the Internet was not given much prominence and, in fact, was mentioned on only 9 pages. Indeed, new products appear almost daily, and even some "standards" in the industry seem to have an extremely short life span. Companies in IT sometimes experience a rapid rise and fall; for example, DEC has obviously lost its leading position in the computer industry, while newer companies such as Google (Mountain View, California) experienced a meteoric rise and became part of our life and culture within only a few years.

Therefore, to avoid the risk of information becoming obsolete too quickly, this MOP will not focus on specific software or hardware products but will instead discuss the general IT management practices that stand the test of time a bit more gracefully. In terms of technical issues, this MOP will stay mostly at the conceptual level and avoid venturing too deeply into the details of specific approaches, technologies, or solutions. However, more specific technical detail may be presented within some of the case studies.

Information technology needs to support business functions and provide benefits to the organization; utilities turn to IT to reduce risks and improve performance, and they expect IT to accomplish specific business objectives. Many business processes within water/wastewater utilities are supported by IT, and some business processes are heavily dependent on IT. Information technology has been broadly applied across different business areas. According to a 1997 publication by the American Water Works Association Research Foundation (Denver, Colorado), *The Utility Business Architecture: Designing for Change,* IT has been used in the following applications:

- Production and delivery of potable water: source protection, water production, water transmission, water treatment, and water distribution.
- Collection and treatment of wastewater: environmental monitoring, solids disposition, wastewater treatment, and wastewater collection.
- Other services: asset and maintenance management; laboratory services; customer service; marketing; developing business plans and strategy; maintaining external and stakeholder relations; planning, designing, and construction of facilities and infrastructure; fleet management; finance and accounting; management and development of human resources; IT management; procurement and inventory management; regulatory compliance; security and emergency response; risk management; land management; and performance management.

Efforts to implement IT solutions in utilities and to execute IT projects have not gone without their share of challenges and difficulties. In some—or perhaps many—organizations, IT projects became infamous for being "always late and always over budget"; there was also a sense that IT did not always fully deliver on its promises. It would be impossible to publish a document that would provide all the information that is required to implement an IT project successfully, to achieve the benefits, and protect a manager or a technical person in a utility from the risks that are inherent in all aspects of IT (e.g., planning, design, development, procurement, implementation, etc.). Even the most powerful and successful computer companies occasionally come out with products that fail to meet needs and expectations (e.g., Microsoft Bob, Microsoft Windows® Vista); in addition, a typical IT department within a water/wastewater utility will have considerably fewer resources than companies that are leaders in the computer industry.

The goal of this MOP is to present an overview of technology that is most relevant to utilities and to provide a reference and a guide that will aid utility managers and staff faced with practical IT issues in their organizations. The MOP is intended to be a document that is broad, addresses most aspects of IT within water and wastewater utilities, and provides some guidance and references to sources that address specific issues in more detail.

Chapter 2 provides an overview of IT within a typical water and/or wastewater utility. It includes brief descriptions of systems and applications and describes a typical "IT landscape" that can be found in a utility. This chapter discusses business drivers for IT in water/wastewater as well as IT applications including business systems, planning systems (e.g., geographic information systems [GIS]), mathematical models, IT systems that support real time operations (supervisory control and data acquisition [SCADA] and process control), and laboratory information management systems.

Chapter 3 addresses issues related to planning for IT within the context of a water/wastewater utility. This chapter describes the methodology and practices for developing a strategic IT plan.

Once a strategic IT plan is in place, an organization needs to develop an IT program. Such a program includes a broader framework for many specific IT projects as well as the scope and schedule for specific projects that will be included in the program. Development of an IT program is covered in Chapter 4.

Chapter 5 addresses issues related to the management of IT projects. Because water/wastewater utility operators must build, operate, and maintain a large physical infrastructure, their approach to project management through most of the history of utilities has been shaped by extensive experience with traditional ("brick and mortar") engineering projects. Information technology projects demand different methodologies and approaches; Chapter 5 addresses project management methodologies that are specifically designed for IT.

Chapter 6 provides a more detailed description of specific components that are part of the IT project and its infrastructure. Chapter 7 addresses issues related to IT security. Because IT systems enable or actually control critical aspects of a utility's business functions, it is important to understand how to manage the vulnerability of IT systems.

Organizational issues, a critical aspect of IT, are described in Chapter 8. Chapter 9 provides a brief summary of critical success factors as well as key challenges in water and wastewater utility projects. This chapter, which was written jointly by all authors of this MOP, identifies present and future challenges in the application of IT within water/wastewater utilities. Finally, Chapter 10 presents examples and case studies to illustrate the concepts discussed in the MOP.

2.0 SMALL VERSUS LARGE UTILITIES

The U.S. Census from 2000 (http://www.census.gov/geo/www/gazetteer/places2k.html) lists 3,219 counties and approximately 18,000 cities and villages in the United States. The water and wastewater utilities that serve these communities are as diverse as the communities themselves. The largest urban centers in the United States include 273 cities with a population greater than 100,000 (http://en.wikipedia.org/wiki/List_of_United_States_cities_by_population). Even within this set of the largest cities, there are significant differences in size. On one end of the spectrum, there are metropolitan areas with large utilities; for example, Metropolitan Water District of Southern California (MWD) (Los Angeles, California) provides water for roughly 18 million people in Southern California. The IT section of MWD includes a staff of more than 150. Two hundred and seventy-third on the list is Wilmington, North Carolina, with a population of 100,192. Water and wastewater for Wilmington is handled by the Cape Fear Public Utility Authority, a utility whose jurisdiction includes the City of Wilmington and surrounding New Hanover County. The Cape Fear Public Utility Authority's IT department includes the IT manager and six staff (network administrator, help desk and server administrator, help desk technician, SCADA administrator, GIS specialist, and one IT analyst). On the small-community end of the spectrum, there are a number of U.S. utilities that serve less than 10,000 customers.

The authors of this MOP tried to present a comprehensive description of different IT systems, issues, and levels of complexity. To achieve this, the MOP contains some content that is primarily applicable to larger utilities that have more complex IT environments. Such large IT environments have more components and a larger IT "footprint." Thus, one can see a more complete picture of how IT could be structured into a comprehensive solution that addresses issues related to managing a water/wastewater utility. Large utilities typically have more elaborate IT environments, bigger projects, more staff, larger budgets, and more complex systems. To some readers with a small-utility perspective, it may appear that there is a bias toward larger utilities. This was not the intent of the authors; however, it was not possible to create an MOP where all content was applicable to all utilities.

Size is not the sole determining factor in properly structured IT solutions. For example, stricter regulatory requirements might push even a smaller utility toward automation and IT. The best approach for a specific utility will need to take into account the specific business needs that the utility must address. For that reason, it was the intent of the authors to make readers aware of the breadth of issues related to IT at utilities, recognizing that implementation of all of the concepts and functionality described will not necessarily be applicable to all utilities.

To apply the material from this MOP to small utilities, the reader may need to "scale down" consideration of some of the content to better address his or her specific issues. Some of the solutions discussed in this MOP may not be directly applicable as described herein. For example, perhaps a small agency will not need a complex and sophisticated enterprise resource planning system and can appropriately use simpler function-specific tools to manage the business. However, there may be lessons learned from larger systems, and some aspects of larger systems can be incorporated on a smaller scale. Even a smaller system will need to select solutions that are based on user requirements, and users across all utilities have similar requirements such as generating and tracking work orders. Therefore, the generic information presented in this MOP can be useful, even though it may not always be directly applicable to smaller utilities.

For the reader who is primarily interested in small utilities, the following are some additional suggestions for easier navigation through this MOP:

- Chapter 2 primarily discusses the variety of software systems typically found in larger utilities. The solution for a small utility will need to be scaled down and adjusted based on the number of customers, users, size of infrastructure, and budget. However, the core functionality of the systems and their role in the utility's business will share many commonalities regardless of the utility's size.

- The scale of IT planning efforts and projects discussed in Chapters 3 and 4 is also based on larger utilities. These are important activities that should be completed by all utilities. The variation between large and small utilities will be the amount of time and number of staff involved in the efforts. Although the implementation of these ideas will vary in their extent, the principles apply to any utility making IT investments.
- The preceding is also true of the content in Chapters 5, 8, and 9. The reader will again need to consider how to scale down the concepts presented when it comes to specific recommendations or solutions.
- Chapters 6 and 7, in particular, contain a substantial amount of complex technical content and, on the surface, may appear to be applicable only to large utilities. However, as previously noted, it is the authors' intent to inform readers about the range of considerations, and there are important lessons to be learned from utilities of all sizes in terms of user requirements, business modeling, infrastructure, security, and procurement.

3.0 REFERENCES

American Water Works Association Research Foundation (1997) *The Utility Business Architecture: Designing for Change*, AWWARF 90726; AWWA Research Foundation: Denver, Colorado.

Gates, B. (1995) *The Road Ahead;* Viking Penguin: New York.

4.0 SUGGESTED READINGS

Drucker, P. F.; Garvin, D.; Leonard, D.; Straus, S.; Brown, J. S. (1998) *Harvard Business Review on Knowledge Management;* Harvard Business School Press: Cambridge, Massachusetts.

Fowler, M.; Scott, K. (2000) *UML Distilled*, 2nd ed.; Addison-Wesley: Reading, Massachusetts.

Hammer, M. (1997) *Beyond Reengineering: How the Process-Centered Organization Is Changing Our Work and Our Lives;* HarperBusiness: New York.

Jacobson, I.; Ericsson, M.; Jacobson, A. (1994) *The Object Advantage—Business Process Re-Engineering with Object Technology;* Addison-Wesley: Reading, Massachusetts.

Jentgen, L. A.; Conrad, S.; Kidder, H.; Lee, T.; Barnett, M.; Woolschlager, J. (2005) Optimizing Operations at JEA's Water System; Project #3001; American Water Works Association Research Foundation: Denver, Colorado.

Kruchten, P. (1999) *The Rational Unified Process: An Introduction;* Addison-Wesley: Reading, Massachusetts.

Paulk, M. C.; Weber, C. V.; Curtis, B.; Chrissis, M. B., Eds. (1995) *The Capability Maturity Model: Guidelines for Improving the Software Process;* Addison-Wesley: Reading, Massachusetts.

Porter, M. E. (1985) *Competitive Advantage: Creating and Sustaining Superior Performance;* Free Press: New York.

Chapter 2

Information Technology in Water and Wastewater Utilities

(continued)

	Management	Finance	Procurement	HR	Customer service	Operations	Maintenance	Treatment	Engineering	Meters	EHS	IT
Accounting/GL	X	X										
Purchasing/procurement	X	X	X			X	X	X	X	X		X
Accounts payable		X	X									
Accounts peceivable		X			X							
CIS/billing		X			X					X		
AMR					X					X		
Inventory		X	X			X	X			X		X
Staff/payroll		X		X								
Timekeeping	X	X		X		X	X				X	X
Work order management	X	X			X	X	X			X	X	
Fleet management	X	X	X			X	X			X	X	
Maintenance management	X											
Asset management	X	X				X	X		X			
DAQ						X		X				
SCADA						X		X				
Security/access control	X			X	X	X	X	X			X	X
Process control						X		X				
Contaminant monitoring					X	X	X	X			X	
CADD									X			
GIS	X				X	X	X		X		X	
LIMS						X		X				
Document management	X	X	X	X	X	X	X	X	X	X	X	X
Mathematical models						X	X	X	X			

FIGURE 2.1 Software tools and users.

The intent of this chapter is to provide an overview of some of the most commonly used software applications in water and wastewater utilities to help utility professionals understand the fundamentals of different systems. This includes the typical end users, data, outputs, and basic functionality for each system. The matrix in Figure 2.1 provides a general overview of the primary users by department as well as type of software system.

1.0 BUSINESS DRIVERS FOR INFORMATION TECHNOLOGY IN WATER AND WASTEWATER UTILITIES

1.1 Historical Perspective

1.1.1 Developments in Utilities

The water industry in the United States, which accounts for approximately 2% of the gross domestic product, has greater control over the natural environment than other industries (Griggs, 1996). The global market for water utilities and wastewater treatment increased at an average annual rate of 3% from 1996 to 2001, reaching

$160.8 billion. Developed countries account for the vast majority of this market. In 2000, the United States accounted for 43% of the global market for wastewater treatment and water utilities (Baumert and Bloodgood, 2004). From 1994 to 2002, revenues earned by the U.S. wastewater treatment and water utilities industry increased at an average annual rate of 3%, reaching $61.2 billion in 2004 (Baumert and Bloodgood, 2004).

In the past, it had not been standard practice for many water utilities to collect sufficient data or to perform the studies necessary to properly understand the patterns and levels of water used by their customers. The end result has typically been large, infrequent, capital-intensive water supply projects interspersed with long periods of excess capacity. In many parts of the United States, water and wastewater utilities operating today face numerous challenges such as rising capital costs for aging water infrastructure renovations, needs for new development, security upgrades, changes in global climate, and more stringent government regulations.

Implementing innovative information technology (IT) solutions and best practices that can facilitate critical business processes (i.e., asset management, knowledge management, and resource management) can help water and wastewater utilities address some of these challenges. However, developing and applying new technology and associated best practices in collecting, storing, and managing water and wastewater information requires expertise, a strong incentive system, sufficient research and development funding, and adequate operating funds.

1.1.2 Developments in Information Technology

Information technology in water and wastewater utilities continues to progress from solely providing operational support toward serving as an enabler for a larger variety of business challenges. Information technology, in essence, is constantly evolving, thereby making it difficult to manage. New developments are steadily replacing or enhancing previous innovations.

There are two trends in the application of technology to supporting utilities. The first has to do with traditionally "unintelligent" assets such as pumps, pipes, and vehicles becoming instrumented, interconnected, and "intelligent" through the use of onboard IT systems. This opens up many possibilities, evidenced by things such as an intelligent utility network that IBM Corporation (Costa Mesa, California) has designed to tune maintenance programs based on highly accurate and timely asset performance information. The application of a sophisticated smart network, however, opens up many new management challenges for utilities in terms of redefining maintenance needs and the staff skills required to provide maintenance for the technology vs physical asset maintenance that is currently done. By educating the workforce and clearly defining roles and responsibilities associated with the use of technology, utilities will be able to achieve the potential value of technology for the organization.

The second trend has to do with the consolidation of business processes and supporting IT systems. An area that is receiving increasing attention in terms of the value of applying technology to support utility decision making and operations is asset management. From utilities across the United States to military bases, the desire to more effectively and efficiently manage a utility's assets is putting demands on utilities to transform from largely paper-based processes to computerized asset life-cycle management that involves the synchronization or direct integration of multiple information systems, including computerized maintenance management systems (CMMS), geographic information systems (GIS), computer-aided design, financial systems, and content management systems.

Over time, the use of technology to support management will evolve and increase. This will include the use of technology integrated with assets such as self-diagnostic and radio frequency identification (RFID) chips that communicate status, problems, and performance metrics directly and in real time to operations management. Permanent assets, such as pipelines, will be able to monitor and report back their capacity, use, and downtimes.

1.2 Business Drivers

Developing IT and its creative application to water and wastewater utilities' business units can help utilities achieve increasing levels of productivity and efficiency. This section outlines some of the business drivers for water and wastewater utilities to implement advanced IT solutions.

1.2.1 *Customer Service*

As the public becomes more technically sophisticated, they expect the organizations they do business with to have complete and easy access to customer information. Faster response times to customer complaints and claims, improved customer service, round-the-clock self service, and improved reliability are among the demands of water and wastewater customers. Information technology-enabled customer service can help utilities meet these demands by making data more easily accessible to support customer inquiries and needs. Information technology can help utility staff with easy and fast access to timely accurate customer information for quality decision making. These systems can provide data to support strategic and capital planning and monitor levels of service. This can then lead to increased customer satisfaction and call center efficiency using an integrated customer service collaboration solution that automates standard services and provides fast access and a single view into all data and work history related to customer locations, including water consumption, quality, service disruption, speed of call resolution, and volume of calls.

1.2.2 *Efficiency*

Dramatic increases in needs and costs challenge water and wastewater utilities. Developing appropriate standard policies and processes and implementing complementary IT solutions in support of these policies and processes can help utilities become more efficient in operations and increase productivity and quality. Examples of this can be seen in utilities' implementations of field technology and asset management programs (Ramon and Stern, 2004).

An example of an advanced IT solution is the implementation of a service-oriented architecture, which lets utilities develop and use best practice templates and creates new applications on top of existing ones, thus increasing access to information for broader uses and thereby increasing the value of the data in current systems. Some examples of IT solutions are online collaboration and mobile technology. Collaboration technology greatly improves the handling of routine jobs and reduces administration burdens and electronic forms that replace paper-based ones. Coupled with updated business processes that address information collection, mobile and RFID, and GIS, IT can deliver up-to-the-minute visibility into the placement, use, and condition of assets in the field.

1.2.3 *Security*

Damage to, or destruction of, the nation's water supply and water quality infrastructure by man-made or natural disasters could disrupt the delivery of vital human services and threaten public health and the environment. Besides physical security, data and technology security are among the important considerations for today's utilities. Early warning systems with real-time monitoring sensors and emergency response systems, customized to immediately alert key personnel regarding any security problems, are some examples of IT that could help utilities mitigate and manage security risks.

1.2.4 *Regulatory*

Replacing or upgrading water and wastewater systems, meeting rising demand, and accommodating rising water quality standards are estimated to range from $300 billion to $1 trillion over the next 20 years (Segal, 2003). Regulatory agencies are taking an increased role in driving infrastructure modernization and, as a result, new regulatory reporting requirements are emerging. Information technology solutions that help utilities with timely and accurate periodic reporting and compliance monitoring could reduce the burden of this reporting. Automated report scheduling and distribution, built-in and customizable reports, and computerized and Web-based permitting are some examples of how IT assists utilities with regulatory requirements.

1.2.5 *Public Trust*

The use of Internet, broadcasting, and alternative media for public relations and for gaining and maintaining public trust is a common phenomenon in today's world. Water and wastewater utilities can increase public awareness of services available, provide information and operations transparency, broadcast audio/video of meetings over the Internet, and publish real-time results on the Web to enhance and maintain public relations and gain public trust. In addition, self-service technology helps customers gain more confidence using the Web and allows utilities to transfer higher-level services to both the Internet and automatic voice services.

1.2.6 *Sustainability*

Global climate change affects drinking water quantity and quality around the world. Global warming may adversely affect water distribution, availability, and quality. Current approaches to resource management are often unsustainable as judged by ecological, economic, and social criteria. Water needs seem more profound today in some regions of the world as availability of fresh water declines, global populations and demands grow, and burdens on waterways increase. Operational sustainability and responses to climate change and natural resource availability are among utilities' challenges today. As utilities look for new ways to reduce effects on the environment, IT plays an important role in working with end users to facilitate the timely collection and use of data needed to calculate and track sustainability related metrics. This could include initiatives such as the development of a "dashboard" to track water and energy use based on data from existing systems and more advanced energy use metering and process control projects to improve energy efficiency.

1.2.7 *Aging Workforce and Changing Demographics*

Many utilities are in the midst of a significant demographic change and will likely see many retirements among management ranks and staff. Fewer young staff entering the workforce compounds this problem. The dual impact of impeding retirements and fewer choices to fill vacancies is a long-term issue. Technology implementation and use may help utilities to further implement technical solutions to address staffing issues, enhance work quality and efficiency, and project the image of a high-tech workforce to attract a prospective, talented, and qualified young workforce. Information technology can facilitate knowledge management that would help diffusion of knowledge from senior to junior staff.

Innovation, teamwork, and professional development can be encouraged and facilitated through the use of IT. Leveraging the knowledge and experiences of existing staff to participate in initiatives to streamline and simplify workforce processes can be mutually beneficial for the utility and employees alike, with the proper outreach and incentives. For example, use of technology in the field to capture information as well as to provide important references to as-built conditions on-demand can enhance worker safety and reduce work downtime. Another example of the application of technology to support the most effective involvement of workers is through workforce roll-out based on skills and availability and innovative employee life-cycle management that aligns employee talents with corporate goals. This will help revitalize the knowledge base being lost as older workers retire.

1.3 Future Business Needs

As a result of the increased speed of communication and availability of data and technology, changes in the global economy and climate, growing populations, and increased infrastructure demand, water and wastewater utilities will be required to improve operational and business practices. Solutions connecting project and work management capabilities with scheduling, outage management, and construction planning and solutions integrating core back-office functions and information on customers, meters, hydrants, financials, and the workforce will be in demand. Integrated financial information using standardized processes, providing visibility on capital, operational and third-party expenditures, monitoring project costs and regulatory risks, and integrating business performance information with management processes could all be achieved by the use of IT.

2.0 INFORMATION TECHNOLOGY SYSTEMS AND APPLICATIONS WITHIN WATER AND WASTEWATER UTILITIES

All of the systems discussed in this section are important to running a utility. But how do you know your utility's specific needs for IT systems and applications? The answer is that your IT needs are directly related to your business needs. To determine what, if any, changes need to be made to your current operations, you must first embark on a needs assessment, which is part of strategic planning covered in Chapter 3 of this manual. Depending on the current suite of systems already in use at a utility, the strategic IT planning process may be applied broadly to the entire utility and potential software systems or to a subset of business processes and specific software systems for those processes.

In evaluating and choosing software, there are a number of common terms associated with the type of tool. These are *legacy*, *custom*, and *commercial* systems; *application service provider* (ASP); and *software-as-a-service* (SAAS) options. These terms can be applied to any of the systems discussed in this chapter.

Legacy is used to describe a software system that is older and is typically based on a technology platform that is no longer supported or that was developed with an outdated programming language. This type of system may be highly successful in meeting the business needs of a utility; however, the system poses a business risk in that if it fails it may not be economically or technically feasible to correct the problem, thereby resulting in a gap in functionality. The risk to the utility depends on the type of system and its role in supporting the business.

Custom is a term used to describe an application that is built specifically for an organization based on its functional needs. Section 4.0 of Chapter 6 discusses custom application development in more detail. This approach is typically taken when no commercial product is available to meet business needs. Although custom systems are appropriate in some circumstances, they can be expensive to develop and risky to maintain over the lifetime of the tool.

Commercial off-the-shelf (COTS) is a term for software that is developed and sold as a product by a company that is in the software business. All of the systems discussed in this chapter are available as COTS solutions. Each software company designs, develops, and maintains a tool or set of tools based on their interpretation of the functional needs of the market. Most vendors also support user groups and hold user conferences to gather feedback from their customers on business needs and future software enhancements. When a COTS solution is purchased, the utility typically pays based on the number of end users. There is also a maintenance fee that is charged that is typically 10 to 20% of the initial software purchase price. This fee includes software maintenance and upgrades as well as end-user technical support.

Commercial off-the-shelf software may also be available from the vendor through an ASP or SAAS option. Through a Web-based delivery mechanism, often called "cloud computing," it is possible to access data and applications stored on remote hardware by way of the Internet instead of keeping it all in a local workstation. The benefits of such delivery are more options in the end-user access device, such as a smartphone, a stripped-down netbook, or even an e-book reader. The user can be anywhere, and so can the source for data and applications.

The cloud delivery approach adds to the flexibility to scale bandwidth up or down at will as well as the affordability of pay-as-you-go service, and subtracts energy-devouring hardware from the local environment. Considerations for such a mechanism include trust in your selected vendor to ensure security and business continuity in the event of a system failure. Selection of this approach involves a contractual arrangement with the software vendor in which the vendor provides the hardware and infrastructure to host the software as well as all of the services for software maintenance. This may be a cost-effective option for many utilities, especially those without an IT department or with limited resources to provide software and architecture support. Fees for this service are typically paid monthly or annually based on the number of end users.

There are many business issues to consider regarding utility software ownership and maintenance. In addition to the ASP or SAAS options previously noted, other possibilities include regional utility partnerships and municipal partnerships. While there can be some economic advantages to partnering with other utilities and/or government entities in acquiring new software, there needs to be sufficient similarities in business needs and business processes to select a tool that meets the needs of multiple organizations. The appropriate approach for any specific utility should be considered as part of the strategic planning process, as discussed in Chapter 6, and organizational issues, as discussed in Chapter 8.

Software, itself, is only one element of a software project. Proper use and acceptance of the tool by staff is as important, if not more important. Figure 2.2 shows the three interrelated elements of any software project: business processes, technology, and people. It has already been noted that technology should be selected to fit specific business process needs; this topic is explored in greater detail in subsequent chapters.

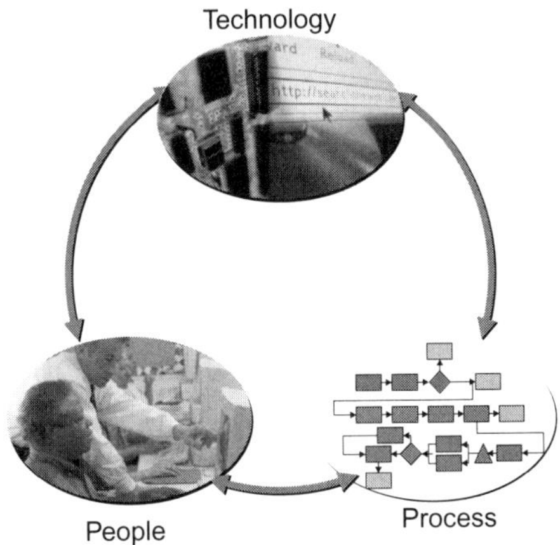

FIGURE 2.2 Key elements of software projects.

The "people" aspects of technology implementation include considerations such as staff involvement in the definition of requirements and software selection, but must carry through to implementation and ongoing use. Training is typically one of the most underfunded and under-recognized critical success factors of any software implementation. While most COTS vendors provide training on the mechanics of using their software, additional training is often beneficial to educate end users on the role technology plays in performing their business processes, especially if these processes will change because of the introduction of a new software or technology tool. End-user and software support documentation is also critical to successfully maintain software over time, to provide a way for staff to explore lesser used functionality, and to provide information to new staff that come on as end users after the initial implementation.

The following sections provide information on a variety of different, commonly used software tools for a utility. Information provided includes a summary of the defining functions of the tool, typical end users, data inputs, outputs, integration areas, and maintenance considerations. No two systems are identical. Indeed, not only are there differences between custom systems and COTS, but also between different COTS solutions and different implementations of the same vendor COTS solution. These differences occur because each utility has a somewhat different set of needs, business processes, and end users. This should not be considered a problem. Rather, it is due to market demand that vendors enable their software tools to be configured to meet specific organizational needs. Subsequent chapters of this manual will go into more detail about how to plan for, specify, select, and implement the tools needed by your utility.

2.1 Business Systems

2.1.1 *Enterprise Resource Planning Systems*

An enterprise resource planning (ERP) system is an information system with multiple modules of functionality serving all the needs of a business. The types of functionality for a utility that are typically included at the core of an ERP implementation include integrated accounting (i.e., general ledger, purchasing, accounts payable, and accounts receivable), human resources, and payroll functionality with configurable workflow and reporting capabilities. Extending ERP can also include the integration of additional data, workflow, and functions such as asset management, work orders, and inventory as well as modules that address customer service, billing, and project management. Additional details can be found in the literature, including *Technologies for Government Transformation: ERP Systems and Beyond* (Kavanagh and Miranda, 2005).

Examples of well-known vendors of ERP systems include Oracle Corporation (Redwood Shores, California), SAP (Newtown Square, Pennsylvania), and Microsoft Corporation (Redmond, Washington) (i.e., Microsoft Dynamics GP, formerly Great Plains software). There are also a number of small- and mid-market vendors that offer this functionality targeted at smaller organizations requiring less complexity in their systems. The vendor marketplace is quite dynamic, and companies are frequently acquiring new tools and phasing out older technologies. Company mergers even make it possible for one company to own multiple products that address similar functionality. This can make navigating the COTS market challenging; as such, it is important to seek up-to-date information on vendors to make the best informed decision about which vendor and product is appropriate.

Descriptions of the functionality of various modules are explained in greater detail in the following sections. However, it is worth noting the benefits and drawbacks of taking an enterprise approach. Enterprise resource planning is a widely used approach to organizational information management and operations. An alternate approach is to address a business's functional needs with a suite of distinct software systems that are synchronized or interfaced with each other as necessary, either directly through the software or through business processes. The idea behind an enterprise approach is to enhance operational efficiency by providing management and staff with a unified interface for entering, updating, and accessing information necessary to run a business.

Implementation of an ERP system can be a significant undertaking. For the software modules to enable a seamless flow of information across an organization, it is necessary to redesign business processes from human resources to work management and financial management to ensure that roles, responsibilities, and workflow are in synch with the software's intent. Business process mapping is a required element of an ERP implementation and can be used to highlight the role of technology in work execution, including data ownership to ensure that data are entered by the right person at the right time to ensure data quality. Business process mapping is discussed in more detail in Chapter 4.

Challenges associated with ERP implementation are primarily related to the additional complexity of operational coordination required because staff in distinct business areas may be required to conduct their work differently to better support staff in other business areas. Integration of software functionality through the modules of an ERP requires an integration of, and appreciation for, a broader set of business processes that can be challenging for utility staff. The success of an ERP approach depends, to a great extent, on the culture of the organization and the willingness to commit to a rigorous process of self assessment and change management. The greatest potential benefit of an ERP approach for a utility may be in the area of asset management. Data on all of the organization's assets, such as people, inventory, and cash flow, are housed in a single database in a manner where the data can be used in many ways to support consistency in the variety of decisions that must be made. If an organization is committed to the duration and effort required for ERP implementation, the benefits over the long term for the utility can be significant. However, it is possible for an organization to meet its needs and operate efficiently without an ERP if the proper processes and roles are in place.

2.1.2 *Human Resources and Payroll*

Human resource systems support the main business processes of an organization related to the life cycle of staff management. Typical functions of a system include hiring, payroll, work/time tracking, benefits administration, training records management, and performance tracking. This functionality may be handled through modules of an ERP system or as a stand-alone system.

Payroll and employee data management are primary functions of a human resource system. Payroll calculations are conducted by using employee data about salary and benefits with data about time worked each pay period, which results in the issuance of paychecks or direct deposits to staff. "Time-capture" systems may be part of a human resource system and are designed to capture data about time worked through a time entry interface, or they may be a distinct software tool for timekeeping that is integrated with payroll and human resource functionality in another system. Another source of data for labor hours spent on different activities and projects may be a work management system or ERP module. These systems are also configured to track and administer employee benefits such as insurance, social security, vacation and sick time, and union participation. The level of sophistication of the functionality will vary by software vendor. All ERP vendors offer human resource functionality and there are also vendors that specialize in human resource functionality.

System end users are typically limited to the few people with authority to oversee the staff's personal data; these end users include the human resource director and any staff responsible for supporting payroll and benefits administration. Some systems may enable direct access to an individual's personal human resource record through the system by that individual and/or the employee's manager.

The core of the human resource system is an employee database. This database may include current and historical information about the employee's skills, degrees, training, and roles in the organization as well as salary and benefits. Other inputs include data about time worked and union participation.

Outputs of the human resource system include payroll for the organization as well as data necessary to support benefits administration such as health insurance and retirement funding. Such systems also allow managers to handle reporting to external organizations such as the Occupational, Safety, and Health Administration; unions; and Social Security Administration.

Human resource systems may be integrated with timekeeping and work management systems to capture data about time worked. In addition, they are often integrated with the financial accounting system as employee payroll and benefits are a significant component of cost for the utility. The human resource system can also be integrated with other utility information systems to drive the system security of those systems. For example, roles defined in the human resource system can be used to determine who should have access to other systems, including project management, financial accounting, and customer information systems, where the level of access to the system depends on an employee's role in the organization.

Because of the high potential for multiple integrations with other systems' functionality, software maintenance may be significant if a human resource system is not run as part of an ERP solution. Data maintenance should be built into daily human resource business processes so the timely update of employee information does not become a difficult or time-consuming task. If a human resource system is integrated with other business systems, it is important that there are strong data quality controls and well-defined data ownership responsibilities to maintain data consistency and integrity for employee data across multiple systems.

2.1.3 *Finance and Accounting*

Finance and accounting systems focus on the financial management and reporting aspects of the utility. These systems center around financial transactions that are reconciled and reported regularly (typically monthly) to balance income received against payments made. The heart of a financial system is the general ledger and the accounts that transactions are posted against. Typical transactions include budgeting, purchasing, accounts payable, and accounts receivable. When utilities undergo their financial audits annually, this is the system upon which auditors primarily focus.

Direct users of the financial system are the finance department or business operations staff of the utility. This includes staff with responsibility for accounts payable, accounts receivable, and reporting of revenue, expenses, capital assets, and depreciation. Indirect users include all consumers of the utility's financial information such as senior management, board members, public utility commissioners, auditors, and financial institutions.

Data entered into the system include all budgets and financial transactions and the accounts to which they are posted. When setting up a financial system, the structure of the general ledger and the organization of various accounts and account types for the utility are entered. At the beginning of the fiscal year, budgets are entered for all accounts. After this period, financial transactions are entered to the system continuously as they occur.

The outputs of the financial system are reports and statements. These are used by utility management to make business decisions about potential changes required in revenue and expenditures, cash flow, borrowing, and capital planning. In addition, outputs are used for inventory reconciliation, annual financial reporting, and documentation to support changes to billing rates.

The utility's financial system has potential integration points with all of the business systems in a utility, including

- Inventory—purchasing and accounts receivable,
- Customer information systems—revenue from billing and accounts receivable from collections,

- Work orders—work costing,
- Project management—capital project budgeting,
- Asset management—asset valuation, and
- Document management.

As with the human resource system, there is a high potential for multiple overlaps with other systems' functionality as it relates to financial data in those systems. Data maintenance may be significant if a financial system is not run as part of an ERP solution. Data maintenance should be built into daily business processes so that the timely update of financial and cost data does not become a difficult or time-consuming task. Reconciliation between an inventory system and a financial system is a common example of areas where maintenance time is required to keep data in synch between systems. If the system is directly integrated with other business systems and not as part of an ERP solution, there may be maintenance issues to address if software upgrades are required by one of the integrated systems. This could have implications for the configuration and upgrade of the financial system.

A utility may choose to replace a financial system because of obsolescence, modernization efforts, or requirements for additional features. Criteria to consider in selecting a new system include required integration points, sophistication of reporting, and ease of use. Data migration between systems as part of an implementation can be straightforward if the originating system is well managed but can become more complicated if integration points are added and additional general ledger/accounts/features are added that affect the configuration.

2.1.4 *Customer Information Systems*

Customer information systems (CIS) house all information about an organization's customers. Customer information systems provide functions such as customer support, account management, billing, and collections. Most CIS also include historical account information for customers such as payment and usage history. Some systems also include meter inventories and are able to issue service orders for meter and field crews related to customer accounts. When a customer calls to open or close an account or to make an inquiry, the CIS is the primary source of information needed to support that call. The CIS is especially important to the financial operations of a utility because it is the system that generates bills and accounts for billing revenues and delinquencies.

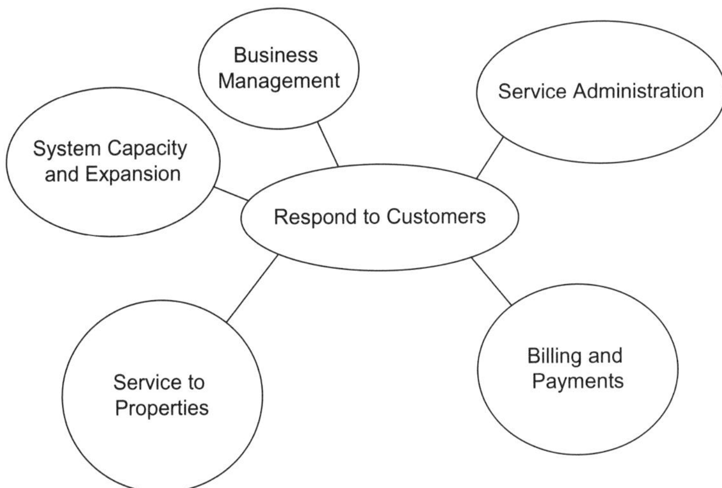

FIGURE 2.3 Customer service-centric view of utility (data from AWWARF [2005]).

Customer service department staff is the primary user of a CIS in terms of inputting data and making inquiries from the system. Figure 2.3 depicts a customer service-oriented perspective of utility operations. The customer relationship management functionality of a CIS is quite useful in supporting a customer-oriented organization by tracking customer complaints and inquiries so that the utility can respond in a timely and appropriate manner. If service order functionality is used in the CIS, the meter department and field crews may also use the CIS to receive and close service orders. If the CIS is set up as a Web-based system to support direct customer inquiries and payments, the actual customers may also be end users with access to only their personal account.

The database of a CIS centers on customers and locations. Each customer is assigned to at least one location. In addition to location, data typically associated with a customer are payment and usage information, including customer rate type, special payment arrangements, and special conditions that may need to be taken into account such as medical conditions that prohibit shutting off service and so on. Location information may include an address, property type, assigned meters, and meter location. Customers and locations are typically updated manually as changes occur. Meter data are typically uploaded according to the customer's billing cycle via an upload from a handheld meter reading device. The CIS will use these data to calculate usage and resulting bill amounts.

A key output of a CIS is customer bill amounts, which are typically generated on a monthly or quarterly basis for each customer. For most utilities, customers are batched into billing routes and cycles that cover most of a month; billing outputs are generated daily. In addition to bills, CIS generates reports related to billing such as high/low meter readings that warrant investigation and past-due accounts that require notification or collection action. If a customer is past due and a decision is being made about whether to shut off service, set up a payment plan, or issue a warning, the business processes associated with making that decision are all supported by CIS.

Other uses of a CIS may be to provide reports of customers to support notification in the event of a service interruption or water quality event. Customer information system data are also used in a number of business operations and planning areas. These include rate studies, water demand calculations, water loss calculations, and water resource planning.

A CIS is a common component of an ERP because of the many overlaps in data between CIS and financials. When not part of an ERP, the CIS is most often integrated with a meter reading system and the financial system. The points of integration with different systems and the CIS are noted in Table 2.1.

Because the CIS system is the heart of the income stream for a utility, it is important that it be technically reliable and that the quality of data be maintained. Data maintenance of a CIS should be done daily. As customer service staff interacts with customers, their account data are typically entered directly to the CIS. It is also useful to periodically run data quality reports to check customer and location data and to set aside time to make necessary corrections. Ongoing data quality checks can greatly improve data quality and make data conversions much easier and faster if there is a migration to a new system.

Replacing a CIS is a significant undertaking that requires dedicated staff time for planning and implementation. For organizations that are running custom legacy

Table 2.1 Points of system integration for a CIS.

System	Integration areas	Purpose of integration
Meter reading	Meter IDs, meter readings	Provide data for billing
Financial system	Billed revenue, accounts receivable	Cash management and financial planning
GIS	Customer locations	Support routing and customer inquiries

systems, the cost of maintaining those systems and providing the level of service expected from customer rate payers often drives the decision to replace the system with a COTS product. When this is done, it is also a good time to assess the organization's meter reading and financial systems as well to ensure that business processes around the aforementioned integration areas are considered in the selection. For organizations using COTS products, a decision may be made to change vendors if the software is no longer supported or if it is found to be inadequate in addressing current functional requirements of the business.

To keep up to date with new functionality, it is recommended that the organization enter into a software maintenance agreement with the vendor. When selecting a vendor, one question that should be asked is how the software upgrade process happens. For some systems, it may be necessary to visit each end-user desktop to apply upgrades; however, this may be undesirable if there are many end users. For more modern Web-based systems, system upgrades and maintenance should be applied to the server only. Ideally, the organization will run a production and test or development instance of the system to help mitigate any unforeseen issues that might arise during an upgrade without affecting business operations.

2.1.5 Document and Content Management Systems

Document and content management systems are also known as *enterprise content management* (ECM) systems. They are used to capture, store, preserve, and deliver any documents or files for an organization. Enterprise content management systems typically have several modules, including document management; collaboration (such as groupware, instant messaging, and video conferencing); Web-content management (Web portals used for publishing information on the Internet through XML, HTML, or browser); records management (for archiving and filing for long-term storage); and workflow/business process management (for standardizing and automating business processes). Enterprise content management systems have several built-in functions to capture data automatically. Among these are optical character recognition, handprint character recognition, intelligent character recognition, optical character recognition, and bar coding. Document imaging functions can be used to both improve the quality of images and to view images. In addition, some ECM systems provide form-capturing functions via scanning. Once the data are captured, then manual indexing or automatic categorization can be used to organize the data.

Typical document management functions include enterprise document search, document check in and check out, version management, file indexing and navigation, visualization, and organization of data through virtual files, folders, and views. By using Web-content management, end users can share information securely over the Internet.

Different kinds of ECM repositories (file systems, content management systems, databases, and data warehouses) can be used in combination for storage. Enterprise content management systems also offer long-term, secure electronic archival of static information. There are various long-term storage media such as storage networks; microfilm; write once, read many, or WORM, optical disks; tapes; and hard disks.

End users of ECM systems cover a broad range depending on an organization's needs. Security and role-based access can be built into these systems and employees of a utility and middle and top management typically would have access to the system as viewers. Typically, view-only users have access to the information based on their roles and have no deleting or editing rights. System administrators are a designated group within an organization with authority to enter new information to the system, organize it, and have rights to edit or delete it. Who should have access to what and who can do what should be agreed on and implemented properly as part of the security planning stage of the project.

All documents relevant to an organization should be captured and stored at a centralized ECM system. These include, but are not limited to, the following: operating procedures, all kinds of reports, manuals, schedules, assessments, standardized forms, plans, drawings, models, surveys, complaints, disputes, pictures, videos, records, and metadata to retrieve and organize data and documents. These inputs are gathered through external sources through manual entry or scanning, database integration with ERP- or CRM-types of systems, intranet data, e-mail, and so forth.

Outputs of an ECM system are typically files, documents and reports that can be shared in various formats such as image file formats (TIFF, JPEG, etc.) PDF, HTML, XML (eXtensible Markup Language), or any other format that is not editable. This information is shared through various distribution channels such as the Web or e-mail, both in digital or hardcopy format.

Depending on the organization's needs enterprise and content management systems could be integrated with (1) an ERP or CRM systems such as SAP or PeopleSoft (Oracle Corporation) for workflow management (where ERP or CRM systems support workflow automation and files are stored in an ECM system); (2) collaboration systems such as email clients, chats, or other forms of solutions that support unstructured contents; and (3) portals such as intranets, extranets, or Web sites for making the portal attractive and more useful.

Integration at infrastructure and platform levels is equally important. For example, the ability to have a single login window for access to multiple applications can become more challenging and costly than it looks. Most ECM tools have database components. Enterprise content management vendors who are not database companies have to balance between supporting enough database formats to attract a broad audience. Similar to any other IT implementation, education and training are the components to success. Many ECM vendors offer training for end-user and IT system administrator staff and often run "train-the-trainer" programs. However, buy-in from top management is critical to help an organization stay focused on getting results and achieving the organization's goals for ECM implementation and use.

Managing content is one of the main challenges of implementing an ECM system due to the range of people at every level of an organization that produce and access content. One of the most important items that must be addressed in an ECM implementation is the definition of the taxonomy to be used. Taxonomies provide the classification structure to enable end users to easily access content stored in the ECM. According to IBM, "...having consistent and reliable access to unstructured content is arguably the foundation to realizing the business benefits of ECM, and all subsequent content-centric enterprise applications will realize their return on investment (ROI) by leveraging this essential capability. As large enterprises standardize on ECM platforms, maintaining the integrity of the catalog is essential to managing access to the volume and heterogeneity of business information..." (Twigg et al., 2007). Some ECM vendors provide tools and services to support the taxonomy identification. It is important that the utility have a good understanding of its content and access needs. During implementation, the taxonomy will be used to configure the ECM tool and guide the migration of content from existing sources to the new tool. For the tool's use and value to be sustainable for the organization, a governance process must also be established to ensure that the taxonomy is applied; otherwise, content retrieval will not meet initial expectations. Migrating existing information to an ECM system requires a big investment in time and labor and the volume of the content might become overwhelming as formatting the original document, identifying metadata for each document, and importing them to the ECM system require following a systematic strategic process. In addition, getting approval for implementing an ECM system can be challenging as the cost of implementing and maintaining these systems typically are not trivial. Although the "people factor" is not related to software and data maintenance considerations, it is important to mention here. Often, end users would be accustomed to storing their files in local hard drives and sharing the content company-wide might become an obstacle as it requires a culture change. With the version control capabilities of an ECM, content creators should be open to making documents more widely accessible to support the benefits of information sharing and collaboration for their team and for the broader organization without worrying about document integrity. In addition, supporting and maintaining an ECM system adds serious responsibilities to the IT department of an organization. Although there are Web-based ECM systems that might be a solution to this challenge, many organizations might choose self-hosted client applications.

2.1.6 *Enterprise Asset Management Systems and Computerized Maintenance Management Systems*

Computerized maintenance management systems (CMMS) have been in existence for many decades, and initially supported plant operations. Technology and business operations at utilities have evolved over the last 15 to 20 years, resulting in the emergence of a suite of functionality that addresses maintenance needs across an enterprise and capitalizes on the data that are part of a typical CMMS to support broader organizational needs for asset data. The term for this system is an *enterprise asset management* (EAM) system. These two systems are generally synonymous in today's software market and will be referred to as *EAM/CMMS* in this chapter.

As Figure 2.4 demonstrates, EAM/CMMS is at the core of a utility's maintenance and asset management activities. Typical functionality includes issuing work orders,

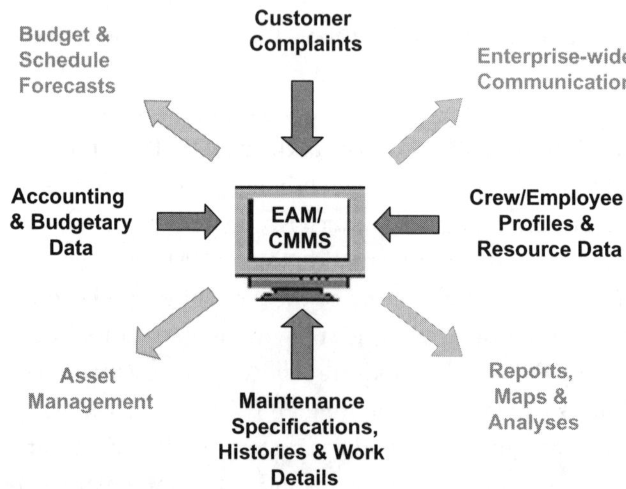

FIGURE 2.4 Role of EAM/CMMS in utility operations.

supporting preventive maintenance, tracking calls, enabling job cost accounting, controlling inventory, and facilitating enterprise asset management. It is also increasingly common for EAM/CMMS tools to have Web-enabled functionality to support remote work and mobile business processes. The core functional areas are described briefly in the following sections.

2.1.6.1 Asset Management

The EAM/CMMS is the core tool for asset management in its ability to track an asset's life-cycle costs (labor and material) and rehabilitations. The data in the system supports optimization of repair versus replacement decisions as well as the development of a capital improvement plan that is based on asset condition and risk.

It is important to recognize that the implementation of an EAM/CMMS alone does not result in an asset management program. However, this tool is at the core of the processes and activities that utility staff must perform in following asset management principles. The general stages of an asset management program include

- Developing an asset register by organizing assets in hierarchy and capturing basic asset attributes,
- Determining the relative criticality of the asset,
- Assessing the condition of the assets and estimating the remaining useful life,
- Assessing the asset risk,
- Developing a maintenance strategy consistent with risk,
- Developing a replacement plan consistent with risk, and
- Developing replacement schedules.

A good EAM/CMMS tool will provide a database structure to capture, store, and report on the data necessary in these activities to implement an asset management program. This includes data elements about assets, including asset attribute data, and descriptive information about condition and history. There are a wide variety of commercial tools on the market that vary in terms of complexity and scalability of the data collected. More sophisticated EAM/CMMS tools may be part of an ERP strategy that includes multiple modules with data and functionality that includes all of the aforementioned functionality as well as fleet management, resource management, and facilities management. Additional references address the topic of implementing asset management programs in more detail (e.g., AWWARF, 2006, 2008).

2.1.6.2 *Call Logging*

As customer service centers receive calls, the EAM/CMMS can be used to create service requests that link to work orders to track problems from initiation to completion. With GIS integration, it can also be possible to geo-locate calls to better support trend identification and work assignments.

2.1.6.3 *Work Orders*

Electronic records for work that need to be completed in the field or at a plant can be issued from the EAM/CMMS. Before completing the work, the system can provide staff with immediate access to asset information and maintenance histories. As work is completed, data about maintenance activities, materials, and labor used are entered to the system to support other business needs. By compiling this information, the EAM/CMMS can then support the analysis of trends in maintenance activities, clustering of issues, and, potentially, cause-and-effect of activities and outcomes. Linking EAM/CMMS to GIS can provide even greater functionality in finding work locations, developing routes and work schedules, and conducting trend analyses.

2.1.6.4 *Preventive Maintenance*

An EAM/CMMS can help move utilities from a reactive response to issues and problems to proactive management of assets and infrastructure. The system is populated with the recommended maintenance schedule for key pieces of equipment and can then automatically generate preventive maintenance work orders based on a defined schedule. This schedule may be based on run times or calendar time. As work is done, it can then track preventive maintenance activities, ultimately providing data necessary to understand trends in asset conditions. The overall result is the minimization of equipment downtime.

2.1.6.5 *Inventory*

An EAM/CMMS can also manage inventory and relate that inventory to work orders. Functionality may include tracking requisitions, automating reordering by setting thresholds and triggers for minimum on-hand quantities, tracking unit price and average price as well as equipment and material usage on work orders. This information can be quite valuable in providing information to support budgeting for operations and capital planning.

2.1.6.6 *Job Cost Accounting*

It is often desirable to be able to identify the total cost of completing a "job" or set of work activities. This may be conducted for accounting reasons as well as to pursue reimbursement from an outside party for work done. The EAM/CMMS uses work order information to aggregate the cost of work by calculating the cost from the labor, material, equipment, and vehicles used on a job. It is also possible to roll up the cost of maintenance for work types and asset types. This functionality is also useful in developing the utility's budgets.

End users of an EAM/CMMS span the organization. Data entry is best handled by those staff with the most direct knowledge of the information being captured. For example, planners and schedulers will enter data necessary to plan and prepare for the work and work crews will document work performed. If a large-scale asset data population effort is being done, a coordinated effort between those staff with the greatest knowledge of the data should be undertaken.

Consumers of information in the system include operations and maintenance staff receiving work orders or inquiring about asset conditions for work that they are going to complete. Additional end users also include customer service and accounting staff and senior management responsible for short- and long-term capital planning, budgeting, and risk management. A comprehensive EAM/CMMS with good data will ultimately result in more coordinated and educated capital decisions.

Additional activities and decisions supported by the outputs of an EAM/CMMS include responding to inquiries about asset locations, work history, and asset conditions. In planning and scheduling work, a system that is integrated with a spatial analysis tool like a GIS will enable geo-locating of maintenance work. At the managerial level, the system will support the development and tracking of key performance indicators that allow for a comparison of productivity statistics both internally and externally.

To achieve the maximum value from an EAM/CMMS solution, there are a number of systems that can be integrated. However, as previously mentioned, it is not the technical system integration that drives the value, it is the coordination of business processes and synchronization of data between systems that ensures the value is achieved. Nontechnical integration with the system listed in Table 2.2 can provide value, even when technical integration is too costly.

As with many of the systems discussed in this chapter, maintenance costs must be considered and planned for in undertaking technological integrations between systems. If one system or module of functionality is replaced with a new system or undergoes a significant upgrade, it can have ripple effects on the other integrated systems. For an EAM/CMMS, there is the additional consideration of identifying the "master" record for asset data. This may be the EAM/CMMS, the GIS, or the financial system, depending on the asset management approach of the utility. Regardless

TABLE 2.2 Points of system integration for an EAM/CMMS.

System	Integration areas	Purpose of integration
GIS	Asset and customer locations	Routing and work scheduling Customer inquiries Geographic trend analysis
Financial system	Asset (value), maintenance budget/costs, and capital budget/costs	Governmental Accounting Standards Board reporting Capital planning Job costing for reimbursement Inventory purchasing
CIS	Customer locations	Link service calls to work orders Call logging
Human resources	Employee data	Job costing of labor resources Work assignment based on staff skills
Operations and maintenance manuals	Equipment servicing guidelines/standards	Preventive maintenance scheduling
Process control	Equipment run times	Preventive maintenance scheduling

of the system of record for asset data, it is critical to have accurate and timely data on assets for the proper operation of a utility. Data entry and data maintenance responsibilities must be clearly defined and regularly completed to support the potentially powerful capabilities of an EAM/CMMS. For example, establishing standard operating procedures and policies for access to operating information, such as process and instrumentation diagram drawings, can ensure the data consistency and quality that is necessary for an EAM system to have maximum value for an organization.

2.1.7 *Project Management Systems*

Projects inevitably generate enormous and complex sets of information. Effectively managing this bulk of information to ensure its availability and accuracy is an important managerial task. Poor or missing information can readily lead to project delays, uneconomical decisions, or even the complete failure of the desired facility. Reports and views are common and crucial requirements as the system is used. Some examples of reports are cash flow; various types of estimates; budgets (current, baseline, or projected) and schedules (Gantt charts and critical path reports); resource allocations; financial commitments; cost to date; spending rates; funding; design and construction status to date; various project status indicators; change order logs; change order reasons; planned vs actual costs; and issues and action items reports. Accuracy and availability of these reports is extremely important for timely decision making and successful project management.

The term, *project management system*, covers a range of software, including scheduling, cost control and budget management, communication and collaboration software, quality management, project documentation or administration software, and resource allocation. Project management systems can be delivered through a variety of system architectures and can range in use and complexity from single user systems running on a single desktop computer to collaborative, multiuser systems that integrate project planning, project control, and management functionality. Project management systems typically come with a built-in reporting system and dashboards for information visualization. It is important to state the differences between a project management system and an enterprise resource management system. An ERP system is implemented and used to manage resources, activities and information for an organization; whereas project management systems are used for managing information, resources and activities for a particular project or sets of projects/programs. Project management systems are utilized for planning and capturing of project information real time; whereas this information becomes part of an ERP system after the project is realized.

End users of a project management system vary depending on the module used and end users' job responsibilities. Users of a project management system include project controls staff (schedulers, cost engineers, estimators, and contract administrators); field staff (inspectors, superintendents, and foremen); project management staff (project engineers and project management); and senior management. The level of detail and information presented change based on the end user's role in a project. For example, a scheduler might want to view tasks, their dependencies, and resources required by each task in a particular project, whereas a project controls manager might want to view schedules of various projects in a master schedule format.

The types of data that are associated with different project management system modules are shown in Table 2.3. The data presented in Table 2.3 demonstrates the overlaps in data used for different uses within a project management system. The modules of functionality do not operate in isolation, and integration between data sets for multiple uses is an important component of such a system.

While project managers implicitly recognize the importance of time and cost information integration for successful project management, it is rare to find effective project control systems that include both elements. Typically, project costs and schedules are recorded and reported by separate application programs. Project

TABLE 2.3 Types of data that are associated with different project management system modules.

Module	Data
Scheduling	Work breakdown structure and tasks Tasks with durations and dependencies Constraints Resources Cost for each activity Milestones
Cost estimating	Work breakdown structure Material quantities and prices Equipment Labor Historical data/cost-estimating reference data Durations
Cost control	Work breakdown structure Budgets Cost accounts Payables Receivables Incurred costs
Contract management	Contract documents Change requests Change orders Amendments Invoices Payment applications
Document control	Project documents, including • Permits • Meeting agenda • Meeting minutes • Correspondence • Disputes • Submittals • Specifications • Memorandums • Regulatory documents

(Continued)

TABLE 2.3 *(Continued)*

Module	Data
	• Drawings • Subcontractor records
Field management	Activity logs Inspection logs Quality control and assurance records Safety and accident records Requests for information Daily field reports

managers must then perform the tedious task of relating the two sets of information. The difficulty of integrating schedule and cost information stems primarily from the level of detail required for effective integration. Typically, a single project activity will involve numerous cost account categories. Similarly, numerous activities might involve expenses associated with particular cost accounts. To integrate cost and schedule information, a common work breakdown structure and specific activities and specific cost accounts must be the basis of analysis.

Another area of integration is between project management systems and ECM or enterprise resource management. Archiving project information when a project is completed is typically a daunting task if done manually rather than automatically. In addition, databases recording the "as-built" geometry and specifications of a facility as well as the subsequent history can be particularly useful during the use and maintenance life-cycle phase of the facility. As changes or repairs are needed, plans for the facility can be accessed from the database.

Project management system vendors offer many types of hosting options, such as the following: use of an ASP, also called *on-demand software* or *SAAS*, where the software vendor provides computer-based services to customers over network and host servers; self-hosting, where the customer hosts servers in-house; or third-party hosting, where customers outsource hosting. Deciding between these hosting options depends on security, retention, storage, bandwidth, and speed needs, as well as the availability of an organization's technical support group. If an ASP model or third-party hosting option is chosen, then, at a minimum, a license should be maintained to access project information or archiving options should be discussed with the service provider. In essence, how to maintain data and access data when a project is completed are important topics to be clarified before investing in any project management system options.

2.2 Planning Systems

2.2.1 *Geographic Information Systems*

A GIS is a database system that is structured to focus on the location and spatial relationships of its date elements. Data are entered, stored, analyzed, managed, and presented in a manner that refers to, or is linked to, location. A utility GIS is an important tool for managing information about the utility's assets and customers. Indeed, wherever location is important or maps are necessary to conduct business for the utility, there is an important role for GIS. The Water Environment Federation's manual of practice (MOP), *GIS Implementation for Water and Wastewater Treatment Facilities*, describes the nine fundamental purposes of a GIS that provide value to users as follows: "mapping and databases, facilities management, facilities atlases, management decision making, facilities planning, federal regulation compliance, business process reengineering, public perception, and e-commerce" (WEF, 2005). The MOP also contains extensive detail on the role and value of a GIS in water and wastewater utilities.

Because of the potential variety and extent of GIS use, there can be end users in many business areas of the utility as well as with stakeholders outside of the utility. Typical end users interact with the data to perform queries and analysis and to create map-based outputs. There is a smaller subset of end users responsible for adding and editing GIS data. To ensure data quality and integrity, it is important to have trained GIS professionals responsible for maintaining system data. Below is a summary of some of the general uses of a GIS for a utility.

Geographic information system professionals are the staff responsible for adding new data to the GIS. The creation and editing of data layers and features in a data layer requires training and experience in the placement of data, the maintenance of metadata, and core geographic principles. The Urban and Regional Information Systems Association (Des Plaines, Illinois) (www.urisa.org) has programs to train GIS professionals; the GIS Certification Institute (Des Plaines, Illinois) (www.gisci.org) also provides certification for GIS professionals.

Field use of a GIS can be beneficial for staff working in the field conducting inspections, maintaining assets, and sampling. It is becoming more common for field staff to carry laptops or handheld devices while conducting their work, and having

Internet access or GIS files on those devices can help them to more accurately and efficiently locate spatial information necessary to complete their tasks.

There is a strong relationship between GIS and computer-aided drafting and design (CADD) software. Computer-aided drafting and design is the most-used software tool in designing utility systems. Once the design is constructed and it becomes an asset for the utility, CADD drawings are often archived as as-built drawings and the data are moved to the GIS. Depending on the organizational structure of the utility, GIS may be part of the engineering department. In many cases, staff may be trained and experienced in CADD and GIS.

At the management level, a GIS can become a key tool to support utility decision making and planning if the GIS is well integrated with other data sources. This can include capital improvement planning and project schedule, public outreach, permitting, and land acquisition. In addition, a GIS is a useful tool when working with other utilities, such as phone, energy, and cable companies, to coordinate road cuts and other project work.

Finally, a GIS can be a powerful tool for communicating with the public. Many governmental entities such as states, cities/towns, and counties now manage extensive GIS data sets and publish them on the Internet for the public to query. There are also private companies, such as Google (Mountain View, California), that provide extensive GIS content through the Internet. Many utilities do not publish their data on the internet, often because of security concerns. However, some layers of information are often used to communicate with the public about large upcoming projects that require public support or that would have an effect on residents. This may be done through GIS capabilities online, such as queries, or through the publication of static maps created with a GIS.

Typical GIS data are structured in layers. Each layer of a GIS represents a different type of feature such as roads, manholes, collection and distribution systems, building footprints, and land cover. Features are represented as points, lines, or polygons. Feature data development can be done by digitizing data or obtaining data from a surveyor or other source of digital information. Tabular data are then created and associated with physical features. This type of tabular data is also called "attribute data." It is attribute data that enable queries to produce maps and reports and perform spatial analysis.

The definition of *specific features* varies depending on the organization. A typical water and wastewater utility GIS includes layers for distribution systems (including pipes, manholes, and plants); collections systems (pipes, catch basins, pumping stations, and plants); connection points with customers; and meters, parcels, land base

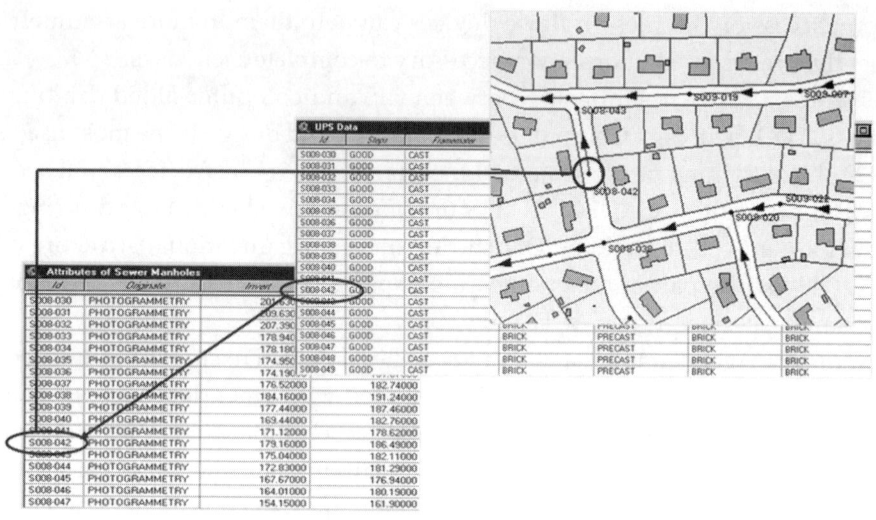

FIGURE 2.5 Spatial data relationships.

and land cover; water features; and infrastructure of other utilities. As depicted in Figure 2.5, each feature may have multiple attributes associated with it, including geographic location, age, material, and owner. Each feature of the same type will have the same attributes; however, attributes will vary by feature type. The common attribute of all features is location.

New layers are typically created through a thorough data development process that includes photogrammetry and development of data collected through aerial photography. The accuracy of the resulting data depends on the quality of the aerial data collected and the procedures applied to create data elements from these data. It is often economically advantageous for utilities to work together or to work with other government agencies in their jurisdiction or neighboring communities to capture and develop GIS data.

Following the initial development of data layers, features and attributes may be added and/or edited. There should be procedures and system security in place to ensure that data edits are done in a manner that ensures data integrity and locational accuracy. In many organizations, data-edit privileges are given to different staff based on their role and the type of data for which they are responsible.

A GIS is best known for its digital map outputs. The combination of digital mapping and database technology is uniquely suited for analyzing objects and areas

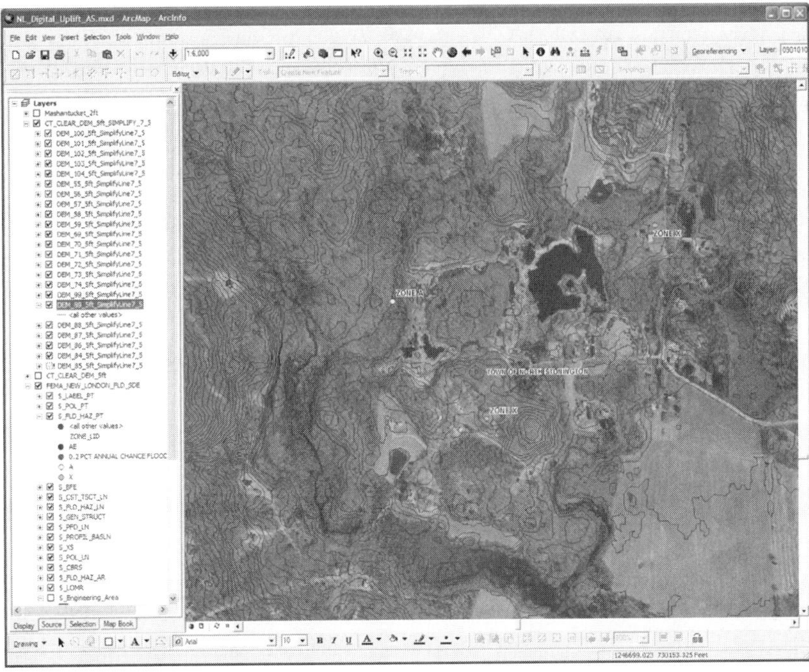

FIGURE 2.6 Multidimensional spatial data.

above, on, or below the surface of the earth and using this technology to generate maps, reports, statistics, and to aid in decision making. With proper planning, data development, and analysis tools, a GIS can be used to support 2-, 3-, and even 4-dimensional analyses, as is shown in Figure 2.6.

The types of outputs from a GIS that are most frequently used in a utility include feature maps of utility assets and customers. These are then used to provide operational support for planning and scheduling capital improvement and maintenance projects, to communicate with the public about utility activities, and to support data modeling activities for predictive purposes.

Geographic information system technology can interact with a number of other technologies that are used in a utility, including CADD software, modeling tools, and field data collection tools like global positioning systems (GPS), as noted in Table 2.4. All of these technology tools can generate data for use in GIS software and can accept GIS data sets. It is also common and effective to integrate GIS with nonspatial data systems in use at a utility.

Table 2.4 Points of system integration for a GIS.

System	GIS integration areas	Purpose of integration
Customer information system	Customer locations	Locate connections Routing service requests Emergency response
CADD	As-built features/assets	Locating assets
Computerized maintenance management/enterprise asset management	Assets	Analysis of asset replacement, repair, and maintenance Routing of work orders
Project management	Project areas	Permitting Land acquisition Health and safety
Modeling	Water features	Water quality Flow modeling

The maintenance of GIS data should be built into the daily business processes of the utility. There are a few different aspects of data maintenance that need to be considered. These are feature data, attribute data, and published data. Feature data should be added regularly as features are added, deleted, or modified in the utility's system using source data such as CADD files from as-built drawings or GPS feature locations collected in the field. Attribute data also require regular maintenance. If the GIS is integrated with other systems, it is possible to automate attribute data updates through synchronization of data sources. For data attributes that are stored directly in the GIS, it is recommended that staff roles be defined to ensure that data maintenance responsibilities are clearly understood and followed.

Within a utility, end users often interact directly with utility data. For security reasons, full utility data sets may not be made available outside of the utility. In some cases, the utility may publish maps for use by stakeholders, including customers, other utilities, and contractors. In some utilities, digital- and paper-based maps may be created to also support field work of their own employees. If outputs are created, it is important that they include a date and that they are updated if and when significant changes occur to ensure that outdated information does not impact decision making or interfere with conducting the business that it was intended to support.

2.2.2 *Mathematical Models*

Mathematical models provide idealized representations of an actual physical system. Models are typically used in a general sense to evaluate the operation of a system or process under actual (typically for troubleshooting) or theoretical (typically for planning) conditions. Well-developed and calibrated models allow a utility to simulate "what if" processes for decision support purposes before attempting changes in the physical system. In summary, mathematical models allow utility stakeholders to both understand how a system is operating now and how it might operate under a different set of conditions. Classic applications include planning for future development (how will my system react to the addition of future customers?), out-of-service scenarios (how can I operate the system with this component removed from service for repairs?), and optimization scenarios (what can I change to better meet a set of operational goals?). Commercial off-the-shelf models are available for all types of system evaluations. The most commonly applied are water distribution models, wastewater and stormwater conveyance (or hydrology/hydraulics) models, and process optimization models (for evaluating treatment processes). Each of these is covered separately in the following sections.

2.2.2.1 *Water Distribution Models*

Water distribution models represent a specific genre of hydraulic model applied to systems operating under pressure. These may include raw water supply, finished water transmission and distribution, manifold force main wastewater systems, and water reuse/delivery systems.

End users of water distribution models are typically limited to engineers, planners, and operations staff. Most models are somewhat complex and, therefore, demand a certain familiarity and repetitive practice to be used effectively. Hence, it is important that staff using hydraulic models be provided both sufficient training and time to accomplish hands-on activities.

Whereas water distribution models can perform numerous functions requiring many types of data, most analyses can be performed by assembling the following six types of data: (1) physical data (diameter, length, roughness coefficient) on the pipes in the system; (2) consumption data that define the demands on the system, that is, the average volume of water to be delivered to each customer; (3) pattern data that define the variability in water demand over the course of a day; (4) elevation data for each feature that enable model calculations of hydraulic grade line to be expressed as a pressure; (5) data for critical features in the system that supply and/or moderate hydraulic grade line, including pumps, tanks, wells, and specialty valves; and (6) operating rules that define how tank levels, pressures, and/or other system parameters will govern the operation of model features, particularly critical features such as pumps.

The most basic water model output includes flow, velocity, and head loss in each pipe and pressure at each junction. These can be for a single point in time (for a steady-state model) or over a time series (for an extended-period model). Models will also provide other results such as system-wide demand, theoretical fire flow availability at any or all node locations, water age, water quality for a conservative or reactive substance, pump energy costs, and so forth. Model output can be used to support utility decisions ranging from how large a replacement pipe should be to the best way to serve an area of low pressure, to how to operate a series of pumps most efficiently to minimize electrical costs.

Due to the variety of data inputs required to enable a functioning model, water distribution models require data from, and can be integrated with, several other systems commonly found in utilities. Table 2.5 illustrates key integration areas.

Software maintenance is typically straightforward, and an annual maintenance fee will enable new versions to be identified automatically and easily downloaded and installed. Data maintenance presents unique challenges for hydraulic models, primarily in the area of GIS integration. This is because the model is not operated directly against enterprise GIS data and also because models tend to be used to

Table 2.5 Key integration areas for water distribution models.

System	Water model integration areas	Purpose of integration
Automated meter reading	Customer demand patterns	Develop individual or representative demand patterns for large users or user classes, respectively
Customer information or billing system	Customer locations and demands	Essential model needed for solving system hydraulics
CADD	As-built features/assets	Can support development of physical network if GIS data are not available or inaccurate
GIS	As-built features/assets	Development of physical network of pipes and junctions
SCADA	Operational aspects	Validate system operations Support development of diurnal demand patterns

evaluate many what-if scenarios that involve the addition of assets that are not present in GIS data. As a result, two-way synchronization schemes are not viable without significant checking and/or customization to avoid unintended consequences (such as proposed model features being deleted because they do not appear in the GIS).

2.2.2.2 *Hydrology and Hydraulics Models*

Hydrology and hydraulics (H&H) models represent a specific genre of hydraulic model applied to systems that are, for the most part, intended to flow under gravity conditions rather than under pressure. These may include open channel/riverine systems, sanitary wastewater collection systems, stormwater collection systems, and/or combined sewer systems. In many respects, H&H models are similar to water distribution models. The following sections, therefore, concentrate on some of the major differences of H&H models.

Like water distribution models, end users of H&H models are typically limited to engineers, planners, and operations staff. As opposed to water distribution models, many H&H models, particularly fully dynamic models, tend to be more subject to instabilities that can require special expertise to understand and correct.

Hydrology and hydraulics models generally require the same types of physical data as water distribution models. One major difference is that H&H models typically place less emphasis on individual customer demands and patterns (an exception is the unusual case where even dry weather flows tax the gravity system) and much more emphasis on rainfall and runoff and/or the effects of rainfall-induced infiltration and inflow. As a result, riverine models require additional data to be collected as well as input related to rainfall and land cover characteristics, especially imperviousness and soil types/characteristics. Closed-system models require information on the levels of rain-dependent infiltration and inflow expected to enter the system. Other differences include the need for cross-section data in riverine systems, the inclusion of support for a variety of pipe shapes (especially in older systems), and the introduction of additional specialty features such as weir and orifice elements.

Hydrology and hydraulics model output primarily includes time series data on flows, velocities, and depth of flow across the system. Like water models, some H&H models can optionally support water quality evaluations such as dissolved oxygen analysis in streams. Model output can be used to support utility decisions ranging from how large a replacement pipe should be to the best way to serve an area of local flooding, to how to operate a series of control structures to minimize combined sewer overflows.

TABLE 2.6 Key integration areas for H&H models.

System	Water model integration areas	Purpose of integration
Automated meter reading	Customer demand patterns	Develop individual or representative demand patterns for large users or user classes, respectively
Customer information or billing system	Customer locations and demands	Needed for establishing base flows in wastewater system models
CADD	As-built features/assets	Can support development of physical network if GIS data are not available or inaccurate
GIS	As-built features/assets	Development of physical network of pipes and junctions Development of areal statistics for model support such as impervious area and soils characteristics
SCADA	Operational aspects	Validate system operations Collect rainfall data

Like water distribution models, H&H models require data from, and can be integrated with, several other systems commonly found in utilities. Table 2.6 illustrates the key integration areas of H&H models. Software and data maintenance requirements are similar to those for water distribution models.

2.2.2.3 Process Models

Process models are commonly used in the water and wastewater industry. A *process model,* in this context, is defined as a mathematical formulation of chemical, biological, or physical processes that occur in water and wastewater process tanks. These models simulate processes that take place in water and wastewater process tanks so that the behavior of full-scale processes can be predicted. The models are implemented via computer software for use in planning, design, and operation of water and wastewater conveyance, storage, and treatment facilities. Process models are used in the planning and design of water and wastewater treatment facilities to size process tanks, to predict removal of pollutants and changes in chemical parameters during treatment such as alkalinity and pH, and to optimize tank geometry. They can be used for operator training, as decision-making guides, or to generate values for control of real-time systems.

Process models may be characterized in several ways. One way to characterize them is to ask whether they attempt to predict what goes on within a process tank by means of a mechanistic method; another way to characterize process models is to ask whether they incorporate probabilistic methods based on prior behavior of full-scale processes. Models like the simulation of single sludge processes (SSSP) model (Clemson University, 1987) of the activated sludge process fall into the former category. Models like the autoregressive-integrated-moving-average (ARlMA) model (Box and Jenkins, 1970) and the Monte Carlo-type model (Metropolis and Ulam, 1949) fall into the latter category.

Another way to characterize process models is based on the degree of knowledge about the underlying flow field that they incorporate. The following three classes of models are identified in this way:

- *Black-box models.* These types of models assume no knowledge about the spatial variation of process parameters within a process tank. Examples of this type of model are residence time distribution theory and the ARlMA model. These models rely entirely on empirical evidence from instances of similar full-scale process configurations for calibration. Their predictive value results from the fact that outputs are typically related to inputs. However, they are disadvantageous compared to models based on the equations of fluid mechanics and biological and chemical kinetics in that they cannot be applied in instances where data from similar applications cannot be obtained. Their predictions may depend on hydraulic residence time in the tank, but make no allowance for differing features of tank geometry.
- *Gray-box models.* This intermediate type of model uses a simplified model of tank flow within the reactor to predict chemical and biological changes within the tank. Examples of this type of model are the dispersion model and the tanks-in-series model. The dispersion model permits flow in one dimension only. It is sometimes called the *dispersed plug-flow model.* Turbulent spreading or dispersion is modeled in either axial or radial directions in pipes and fluidized beds. When dispersion in the axial or longitudinal direction is modeled, it is called the *axial dispersion model.* The tanks-in-series model assumes that transport in a process tank can be modeled by replacement of the real tank by an "equivalent" number of completely mixed single tanks operated in series. This type of model has become widely used for modeling of activated sludge wastewater treatment processes.
- *Glass-box models.* This type of model bases its prediction of tank process behavior on calculation of the multi-dimensional velocity field within the tank. These

models place solids transport or chemical reaction models within a flow field, which has been predicted using mathematical equations of fluid mechanics in two or three dimensions. By predicting the variation in velocity and pollutant concentration within the tank using the equations of computational fluid dynamics (CFD), these types of models can more accurately predict process behavior than the previous two types of models. Models based on CFD have been used for planning and design of sedimentation and disinfection tanks and for analysis of flow problems in treatment plant hydraulic elements. The disadvantage of this type of model is that it can be compute-intensive and requires high-level knowledge of the underlying fluid mechanical theory for effective use.

In general, black-box models are stochastic and gray- and glass-box models are mechanistic, in the sense defined previously.

Process models are typically used today by engineers in planning and design of water and wastewater process facilities or for the diagnosis of problem behaviors in full-scale plants. However, process models can also be effectively used for operator training and for day-to-day decision making in water and wastewater treatment facilities. During the 1990s, the commonwealth of Massachusetts used wastewater treatment simulation software based on black-box models for operator training. Many treatment plants have used dynamic models of plant hydraulics and tanks-in-series models of plant biological processes for day-to-day decision making.

Process models use inputs from water or wastewater treatment plant (WWTP) operation or from projections of future plant operation to simulate future behavior. These data include

- Plant flow;
- Site characteristics such as elevation, air temperature, and wind speed;
- Process fluid characteristics such as temperature and viscosity;
- Process chemical characteristics such as pH and alkalinity;
- Pollutant concentrations such as turbidity, suspended solids, biological and chemical oxygen demand, and ammonia; and
- Tank and conveyance element dimensions.

Process model outputs typically include predictions of effluent characteristics from process tanks such as

- Plant flow;
- Process chemical characteristics such as pH and alkalinity; and
- Process pollutant concentrations such as turbidity, suspended solids, biological and chemical oxygen demand, and ammonia.

Process models are also used to optimize the geometry of process tanks such as sedimentation tanks, disinfection tanks, and hydraulic components. This is done by simulating the behavior of the tanks under different geometric arrangements to predict which geometric features will be most effective during treatment. In this case, the outputs from the process models are comparative profiles of velocity, solids, and/or reactant concentrations within the tank.

Process models can be used to "fill in" missing data for process control of systems such as activated sludge aeration or sludge wasting. They can be included in the dashboard of plant operations personnel to aid in decision making.

Process models are available from a wide variety of sources, including academic, proprietary, commercial, and public domain. Examples of black-box types of models, which have been used in the water and wastewater industry, include neural network models, ARIMA models, and Monte Carlo simulations. Neural network models have been used to forecast inflow into storage tunnels; ARIMA models have been used for evaluation of clarifier performance and for linearization of flow networks. Monte Carlo models have been used for the evaluation of process data for the selection of appropriate peaking factors for wastewater treatment planning. These have all been implemented as proprietary software development projects by individuals or consulting engineering companies.

The SSSP model of the activated sludge process was an early tanks-in-series gray-box type of model. This model is available free of charge from Clemson University (Clemson, South Carolina). The SSSP model was based on the then International Association on Water Pollution Research and Control (IAWPRC) model for single sludge treatment systems. The original IAWPRC model is now part of a series of activated sludge models documented by the now International Water Association (London, United Kingdom) as activated sludge models (ASM) (IWA Task Group, 2000). Proprietary versions of the ASM models have been implemented by a number of individuals and consulting engineering companies. A wide variety of commercial biological process models based on this concept for wastewater treatment simulation are available, including BioWin™ from EnviroSim Associates, Ltd. (Flamborough, Ontario, Canada), WEST™ from MOSTforWATER N.V. (Kortrijk, Belgium), and GPS-X™ from Hydromantis, Inc. (Hamilton, Ontario, Canada).

Glass-box types of models have been implemented using proprietary models, public domain models, or commercial CFD packages. An example of a proprietary model of secondary sedimentation is CLARITY (Vitasovic et al., 1997) . This model used proprietary CFD software to develop simulations of solids profiles in activated sludge sedimentation tanks. More recently, the Q3D and 2DC models were developed at the University of New Orleans (UNO) by students of John A. McCorquodale. Development of these models, which was part of a Ph.D. dissertation at UNO (Griborio, 2004), was partially funded by a grant from the U.S. Environmental Protection Agency (U.S. EPA).

Commercial CFD packages can also be used for modeling of process tanks. These packages are available from a variety of vendors. The software typically includes convenient aids to geometric definition of flow grids, a variety of models for turbulence, automatic generation of contour and other types of plots to enhance visualization of results, and the capability for customization by the addition of a user-defined function. For proprietary models, maintenance of process modeling software is done by in-house programmers and, for commercial products, via updates and licensing agreements with vendors.

2.2.3 *Portals*

An information portal is a technological framework that integrates data from multiple sources to provide end users with a unified perspective of the information. Through a Web-based interface, summaries of data can be presented in a manner that meets specific, customized business needs that cannot be met with a report from a single information system. The portal screens that users interact with and view are often referred to as "dashboards." Commercial enterprise systems that include multiple modules of data to support a variety of business processes may come with portal functionality that is configurable to the organization's needs. Alternatively, portals may be customized to integrate disparate information silos.

Portals are an effective mechanism to provide outputs of information to end users. The inputs to portals are the databases of organizations that are part of other information systems; data are typically inputted directly through technology integrations rather than entered by end users. The content of the inputs and outputs is configurable or customizable based on the information, metrics, and decision-making needs of the utility.

Dashboards can be developed for many different end users with different responsibilities and perspectives to provide information that is operational, tactical, and strategic. In all of these instances, the idea is that end users are decision makers that need access to information and metrics that enable them to move the organization closer to an agreed-upon set of goals in a timely manner. To do this, it is necessary to have access to data that is high quality, timely, and in line with the organization's metrics and goals.

For example, the dashboard of a portal designed for a utility's operational purposes can support end users that are front-line workers and supervisors responsible for monitoring and optimizing operating processes. It can include diagnostic metrics about flowrates, energy consumption, and water quality that are frequently updated to provide the information needed to best run the plants and associated system components. Figure 2.7 presents an example of an operational dashboard that supports sharing frequently to update operational information across jurisdictions to support emergency preparedness and response.

A dashboard designed for tactical purposes typically targets different supervisors and managers who require self-service access to information with direct navigation

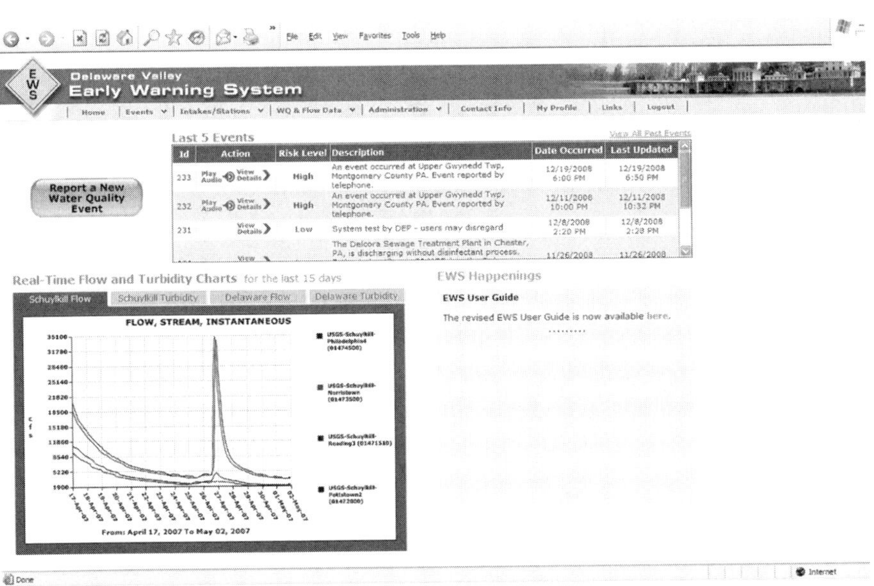

FIGURE 2.7 Delaware Valley Early Warning System home page (courtesy of the Philadelphia Water Dept.)

to monitor progress of metrics against baselines and goals. It can also be possible to enable end users to "drill down" their analyses by targeting a specific metric result and obtaining the details of the calculation behind it. The calculations may reside in the source system or the portal. Data are typically updated less frequently than in an operational system and, depending on business needs, but may be daily or weekly.

Strategic dashboards are developed for use by senior management responsible for measuring high-level objectives and making complex decisions based on qualitative and quantitative information. The information presented typically focuses on longer-term trends, and data may be compiled monthly.

Because portals are, by definition, a mechanism to integrate data sources, the type and number of systems that may be integrated is driven by business needs. "Application access and integration are the bellwether for portal success," with 63% of survey respondents expecting to integrate seven or more systems (Ramos et al., 2004). Because of the complexity of data and system integration, for a portal development project to be successful it is important that sufficient attention be given to the ultimate information and decision-making needs of the organization. Outputs and functionality must be useful and ambitions for integration must be realistic. The integration needs between data sets must be well thought out in addition to the methods and business rules for compilation, calculation, and presentation.

System maintenance considerations for portals depend, to a great extent, on the systems that feed the portal and the technologies applied to develop the portal. For a utility that has established technology standards with common database software and platforms, portal development can be done in a manner that minimizes the maintenance burden. Most maintenance requirements will develop from new user requirements for outputs, changes in business rules, and replacement of the core systems that feed the portal. As with all utility systems, there should be a business owner or set of owners that participate in a regular process to evaluate the portal's ability to meet business needs, and a budget should be set for periodic maintenance to keep the portal current and to maximize its effectiveness for the organization.

2.3 Operational Systems

2.3.1 *Data Acquisition*

Data acquisition is defined as the function of obtaining data from sources external to a microprocessor or computer systems, converting it into binary form, and processing it. Whereas any sensor/actuator combination can respond to immediate status changes, data acquisition is distinguished by the fact that it is based on gathering multiple snapshots or slices of time, analyzing them together, and typically spotting and being able to act on otherwise unseen trends. Data acquisition systems consist of four elements:

(1) Measurement (water and wastewater treatment process sensors, water quality online analyzers);
(2) Recording output signals (logger unit);
(3) Uploading/accessing recorded data (telemetry); and
(4) Analysis of recorded data (data acquisition software).

These four elements have specific requirements that need to be physically present and included in the design process. Sensors must be able to convert any measurement parameter to an electrical signal usable by the recording unit. Sensors to measure selected parameters must meet certain specifications, and the routing of the sensor cables ensures that they will not suffer from electromagnetic interference from other electronic systems. The data acquisition unit (including memory) and the link from the data acquisition unit to the operating platform to upload the acquired data via a hardwire cable or telemetry also must conform to various industry standards as well as end-user requirements. Once data have been collected in the data acquisition and transferred to the utility's computers, safeguards must be in place to protect the data. This may require involvement by the utility's IT group. Questions that must be addressed include the following:

- How will these data be used?
- How long will the data be stored?
- What security measures are in place to protect these data in case of a catastrophe or security breach? (See Chapter 7 for a complete discussion of IT security issues.)
- Who in the utility organization is responsible for maintaining these data?
- What methods will be used to archive the data?
- How much will it cost to process and maintain the data?
- How are software upgrades handled and how often?

2.3.2 *Telemetry*

Telemetry is an automated communications process by which data are collected from instruments located at remote or inaccessible points and transmitted to receiving equipment for measurement, monitoring, display, and recording. Transmission of the information may be over physical pairs of wires, fiber, telecommunication circuits, wireless radios, or satellite. Early telemetry systems used tone-based modulation techniques to transfer analog and digital values at low data rates over telephone lines and radio links. Today, as microprocessor and communications technologies have matured, telemetry systems have largely been replaced by supervisory control and data acquisition (SCADA) systems (Boyer, 2004).

2.3.3 *Supervisory Control and Data Acquisition*

Supervisory control and data acquisition is an industrial measurement and control system consisting of a central host (or master) and one or more remote stations or remote terminal units (RTUs) controlled by standard and/or custom software. Increasingly, programmable logic controllers (PLCs) are being used instead of RTUs in SCADA systems. Programmable logic controllers are very cost effective in such applications and are easily programmed using industry standard programming languages as defined in International Electrotechnical Commission (Geneva, Switzerland) Standard 61131–3 (http://www.plcopen.org/pages/tc1_standards/iec_61131_3/).

The host or master terminal unit (MTU) performs supervisory monitoring and control of the process via RTUs or PLCs. If communication between the MTU or host is lost, both PLCs and RTUs can operate in a stand-alone mode independent of the host or MTU. In PLC-based systems and the newer programmable automation controller (PAC)-based SCADA systems, the host typically resides on a personal computer with human-machine-interface (HMI) software. An HMI is the apparatus that presents process data to a human operator and, through this, the human operator monitors and controls the process.

Programmable automation controllers represent the latest technology used in SCADA systems today. A PAC is a compact controller that combines the features and capabilities of a personal computer-based control system with that of a typical PLC.

Distributed control systems (DCS) are similar to SCADA systems and are routinely seen in factories, treatment plants, and so forth. These systems typically communicate to remote locations over a local area network, or LAN, connection. Supervisory control and data acquisition systems often cover larger geographic areas and rely on a variety of communications systems (i.e., cellular, licensed and unlicensed radios, telephone lines, satellite, etc.). In addition, SCADA systems use a sophisticated database, provide graphing and reporting functions, offer an interface to operate equipment, and have software-initiated alarms to alert plant personnel to specific conditions.

Some examples of telemetry and SCADA applications in water and wastewater treatment include water supply systems, wastewater treatment plants, and sewer and stormwater collection systems.

Supervisory control and data acquisition systems have evolved in that they are able to provide "near real-time" updates from thousands of RTUs, PLCs, or PACs that are often spread across large geographical areas using a range of secure communications media to multiple "users" that may also be remotely located or to a central location. Until recently, SCADA systems were most often used in a reactive manner to identify system faults as they occurred and to record system data and events for later analysis. Present-day demands on all types of businesses, including water and wastewater treatment for increased efficiencies, regulatory agencies for increased oversight, and municipal utilities, in particular, for increased security of their assets, means SCADA systems must now be proactive and include a lot of data management and security functionality that allows problems to be avoided rather than just recorded.

2.3.4 *Process Control*

2.3.4.1 *Description and Functionality*

Process control is a broad term applicable to many industries. In this manual, process control is discussed in the context of water and wastewater treatment. A useful definition of a *process* is as follows: "A sequence of chemical, physical, or biological activities for the conversion, transport, or storage of material or energy" (ISA, 2003). Conversely, *process control* is defined as the "control activity that includes the control functions that are needed to provide sequential, regulatory, and discrete control and to gather and display data" (ISA, 2003). For water and wastewater treatment, this includes equipment, sensors, analyzers, actuators, valves, controllers and computers, and software used in the treatment process.

End users of the water and wastewater treatment process are the utility's customers who benefit from having clean water at their tap and WWTPs. From an equipment manufacturer's perspective, end users represent the utility's management personnel and, from the utility's perspective, the operation and maintenance personnel.

Process control uses computers and software related to SCADA, telemetry, DCS, PLCs, and PACs to perform specific reporting, storing, monitoring, and control tasks. In most instances, a digital control system is used. All of these systems process data and, in turn, perform the control needed to achieve the desired process state or treatment objective.

Digital systems are rapidly changing with technology. Consequently, new terms and acronyms are constantly being developed to describe the technology. The following are some basic terms that can be defined at this time:

- *Input/output*—refers to the information coming into and coming out of the digital control system. There are several input/output types encountered in a typical process control system.
- *Analog input*—any input to a digital system that is continuous in nature. A common example is the 4- to 20-mA signal used to represent a process variable (i.e., temperature, pressure, flow, level, modulating valve position, dissolved oxygen, pH).
- *Analog output*—any output from a digital system that is continuous in nature. A common example is the 4- to 20-mA signal used to modulate a valve or control the speed of a motor.
- *Discrete input*—any input to a digital system that represents one of two states (i.e., on/off, open/close, high/low level).
- *Discrete output*—any output from the digital system that controls a two-state device (i.e., open/close, on/off, raise/lower).

All digital control systems include a combination of hardware and software. A typical PLC- or PAC-based DCS architecture consists of multiple cabinets that house the inputs/outputs, controllers, power supply, communication and other modules, consoles for operator workstations (typically personal computers or panel-mounted industrial operator interface terminals), a data communication link, digital storage devices (optical disk, computer drives, etc.), visualization software (or HMI), and, sometimes, a historian (personal computer type) for long-term data archiving and analysis. To implement a complete integrated digital process control system, all these systems must be properly selected, configured, and installed. For example, DCS configuration requires a combination of fill-in-the-blanks data entry for standard inputs/outputs and control algorithms (i.e., control logic or sequence of operations). Such configuration is done on a personal computer with the manufacturer software and within a database that is common to all of the hardware modules in the system, with the resulting application database downloaded to the respective modules.

Most digital process control systems that use the various software types described previously will need maintenance at some point. This includes updating the operating system (i.e., Microsoft Windows® or other operating systems), antivirus program, security software, firewall management, changes to control algorithms based on process needs, and version updates for HMI software. A utility must have staff dedicated to these tasks and, at a minimum, have staff with basic familiarity of these systems to serve as a liaison with the system supplier or system integrator who can then be contracted for these tasks. An ideal instrumentation, control, and automation (ICA) system contains the following functional components:

- A quality team of people who feel a deep sense of ownership of the system and treatment plant and who are committed to continuous improvement ethics;
- An instrumentation system to gather adequate process variable information;
- A monitoring system to acquire data, process and display the data, detect and isolate abnormal situations, assist in diagnosis and advice, and, finally, simulate the consequences of operational adjustments; proper data acquisition and reporting is crucial;
- A control system to meet the goals of the operation. This can take place both locally within the treatment process by low-level control system or by coordination of the various processes within the plant as well as with the sewer system; and
- A network and peripheral devices to transport and transpose the signals.

2.3.4.2 Incentives for Instrumentation and Control

Advanced control is becoming increasingly demanded in water and wastewater treatment systems. Various case studies have shown significant savings in operating costs and remarkably short payback times. The application of process control in wastewater treatment systems, however, has been developing much later than in the chemical or pulp and paper process industry and, in fact, the water treatment industry can learn from these industries. It is the process knowledge, sensor technology, and the way the plants have been designed and built that may limit what can be achieved today. Wastewater processes do have some unique features, including flowrates, disturbances, small concentrations, organisms, separation, and the fact that all the "raw material" has to be accepted and treated.

The need for ICA was recognized as early as the 1970s as industrial control systems were developed for the process industry. Still, however, dynamical systems and process control are seldom part of the general civil or environmental engineering curriculum. Therefore, many wastewater treatment system designers are unaware of the potential of ICA.

Development of ICA applications in wastewater treatment systems has been driven by the following factors:

- *Instrumentation technology*—"to measure is to know"—has matured and complex instruments like online, in situ nutrient sensors are increasingly being used for control purposes.
- *Actuators* have improved over the years. Variable-speed drives in pumps and compressors represent a proven technology that allows for better control of the plant.
- *Computing power*—many utilities are designing and installing their second and sometimes even third generation of SCADA and process control systems. Benefits of such systems are no longer questioned. Data-processing tools are mostly borrowed from multivariate statistics and soft computing (neural networks and fuzzy systems). Integration of these tools with low-level control loops in the process is yet to be explored.
- *Control theory* and *automation technology* offer powerful tools. Benchmarking of different control methods has been established and some novel tools for evaluating control strategy performance, such as costs, robustness, and "performance images," have been developed.
- Advanced *dynamical models* of many unit processes have been developed and there are commercial *simulators* available to condense the knowledge of plant dynamics.
- *Operators and process engineers* are much more educated in instrumentation, computers, and control ideas today. However, there is still a great need for better education.
- There are obvious *incentives* for ICA, not the least of which is from an economic standpoint. Plants are also becoming increasingly more complex, which necessitates automation and control.

2.3.4.3 Conditions to Be Met in Process Control

Process control is not only about running equipment such as pumps, compressors, and valves. It is also about consistently meeting the requirements of the operation while minimizing operating costs. This means that the plant system has to be understood from a dynamical point of view and the plant owner has to be a competent customer. It is also crucial that operators are part of both the design process and the control system definition.

Measurements of adequate process variables have to be provided with adequate sampling rates. Monitoring is an important part of the system and can provide critical information about the plant as well as present early warning signs of disturbances and process changes. The actuators, such as pumps, valves, and compressors, have to be designed so that the plant is truly controllable. Indeed, too many control systems have failed because of inadequate actuators. Only the team at the plant can provide true improvements to the operation.

The primary inputs to a control system are measurements from online sensors. Many sensors, like dissolved oxygen probes, provide continuous signals. However, the true sampling rate has to be related to the dynamics of the process. For example, it is meaningless to deliver dissolved oxygen data every second because the result of an airflow change will not be significant until 15 to 30 minutes after a control signal change.

Thus, it is critical to first eliminate noise from the measurements and then to supply a filtered signal at an adequate rate. One can see full-scale plants where a dissolved oxygen control system will adjust the airflow every minute. This is completely meaningless, and such a system tries to control the noise instead of the real variable. Furthermore, the equipment can easily get worn out. In this specific case, the sampling rate typically may be once every 6 to 15 minutes.

Unit processes with different dynamics require different kinds of sampling rates and different noise-elimination methods. Concentration dynamics for an aerator are on the order of several hours, while excess sludge pumping control should have a time horizon of many days.

It is crucial that the plant is controllable via pumps, valves, compressors, and other devices. Many plants lack this kind of flexibility and, consequently, control of such a plant will not be successful. There are many kinds of unsuitable actuators. An on/off pump for influent flow will generate unnecessary disturbances in the plant. Such operations will often upset the settler and clarifier operation. Some systems are now provided with compressors that will allow any variable airflow. However, this is energy inefficient.

2.3.4.4 Integrated Operation

Integration aims to minimize the effect on the receiving water while ensuring better resource use. System resilience is an important factor with integration. Specifically, this includes the ability to attenuate disturbances while also being sensitive to significant disturbances or even purposeful and harmful attacks. The ultimate goal of an integrated approach is to formulate some criterion for the receiving water and its ecological quality while satisfying various economical and technical constraints. The challenge lies in relating this performance to plant effluent and potential sewer overflows. Performance measures are needed for plant operations that relate effluent quality to the resources needed to obtain them, such as energy, chemicals and other materials, and operating costs. Models are being developed to find dynamic strategies to maximize a WWTP's loading according to continuous monitoring and prediction of the operational state. One example is maximizing the nitrification capacity in the activated sludge process, depending on the load to the system. Rosen et al. (2005) reported some full-scale results using this model. Another aspect is *storage management* (in the sewer system and in retention tanks), not only during storms but also during normal operations. By mixing different types of wastewater to compensate for nutrient deficit or overload, for example, the capacity of the plant can be maximized.

All integration translates to some kind of compromise. If there were no interactions, then the individual optimization of each subprocess would be the best strategy. In reality, having couplings aims at a better result rather than if each one of the processes were controlled separately. For example, if sewer and wastewater treatment operations are to be integrated, then two competing goals have to be satisfied. The goal of the sewer operation is to minimize the combined sewer overflow. That may mean that the WWTP gets overloaded. Alternatively, the operator of the WWTP wishes to maximize the incoming load. Thus, the integration has to meet both these requirements. Other examples can be found in the operation of activated sludge systems (Olsson and Newell, 1999).

2.3.4.5 Software Considerations

A control system contains some kind of real-time database. This will include all relevant signals from online instruments and manual observations. It is natural that sampling rates of different variables are different and that the quality of the data is also different. Any data value has to be screened and tested. Many different kinds of software modules have to retrieve data from such a database, including control modules, graphical presentation programs for widely variable timeframes, estimators for indirectly calculated variables, and various modules of early warning and detection modules. There are also a variety of operational tools available that can be integrated with control systems to support data mining and/or reporting to facilitate operational decision making using data from SCADA, laboratory information management systems (LIMS), and other similar systems.

A control system has to be viable, which also means that it will be continuously updated. New control loops can be added and, therefore, the database has to allow new modules that can access it.

2.3.4.6 Incentives for Control

Instrumentation, control, and automation have significantly increased the capacity of biological nutrient-removing WWTPs. Advanced knowledge of mechanisms involved in biological nutrient removal that is being gained today is producing an increased understanding of the processes and the possibility to control them. There is a sophisticated relationship between the operational parameters in a treatment system and its microbial population and biochemical reactions, and, hence, its performance. With further understanding and exploitation of these relationships, improvements as a result of ICA may reach another 20 to 50% of total system investments within the next 10 to 20 years.

Advanced control is becoming increasingly in demand for water and wastewater treatment systems. Various case studies have shown significant savings in operating costs and remarkably short payback times. It is process knowledge, sensor technology, and the way plants have been designed and built that may limit what can be achieved today. Wastewater processes do have some unique features compared to other process industries; these are flowrates, disturbances, small concentrations, organisms, separation, and the fact that all the "raw material" has to be accepted and treated. What are really different are the attitudes and incentives in the different industries. Of course, the attitudes often depend on the incentives.

Disturbances in wastewater treatment systems are significant and the main reason for control is explained in this section. If the influence of disturbances can be attenuated, then the plant can be operated consistently 24 hours a day. Instrumentation, control, and automation make it possible to operate the plant unmanned at night and during weekends. Furthermore, ICA contributes to lower operating costs, such as electric power requirements for pumping and aeration, and less chemical consumption. Instrumentation, control, and automation are also increasingly in demand in complex biological nutrient removal plants. On the unit process level, there is a competition between different reactions. Oxygen is a shared common resource for several biological reactions. Carbon is needed for both anaerobic operation and for denitrification. A key feature of ICA systems is to provide online monitoring to detect deviations from good operation. In a plant operation, ICA should be increasingly used to match the operation of various unit processes.

Because of the many recycles in a plant, the various unit processes should not be operated separately. For example, the return sludge flowrate couples the aerator to the settler and sludge recycle from an anaerobic digester makes the coupling to the activated sludge process important.

One incentive for control is the presence of disturbances in a plant. Action needs to be taken to compensate for the effect of disturbances. It is even better if disturbances can be attenuated or even eliminated *before* they hit the plant. Compared to most other process industries, many disturbances that affect a WWTP are extremely large. Discrete events such as rainstorms, toxic spills, and peak loads also occur from time to time. As a result, the plant is hardly ever in steady state, but is subject to transient behavior all the time.

Consistent performance must be maintained in the presence of these disturbances. The traditional way of dampening disturbances has been to design plants with large volumes to attenuate large-load disturbances. This solution incurs large capital costs. Online control systems, which have been demonstrated to cope well with most of these variations, are a much more cost-effective and, thus, attractive alternative. Disturbance rejection is indeed one of the significant incentives for introducing online process control to wastewater treatment systems.

Many disturbances are related to plant influent flow. The influent is changing both in terms of flowrate, concentrations, and composition (Figure 2.8), with time scales ranging between hours to months.

If the result of the disturbance is measured *within* the plant, such as a change in the dissolved oxygen level, a rising sludge blanket, or a varying suspended solids concentration, the measured information is *fed back* to a controller that will activate a pump, a valve, or a compressor so that the influence on the plant behavior is minimized.

Sometimes, a load change can be measured upstream before it has entered the plant. Then, the information can be *fed forward* to prepare the plant. For example, the aeration can be increased before a load increase hits the plant. Another example is when the return sludge pumping can be increased to lower the sludge blanket to prepare the settler for an expected increase of the hydraulic load.

Unfortunately, many disturbances are created within the plant because of inadequate operation. This often depends on a lack of understanding of how the various parts of the plant interact. Figure 2.9 presents such an example. The influent flow is pumped via three on/off pumps. This results in sudden changes in the flowrate that will have a detrimental effect on the behavior of the secondary clarifier.

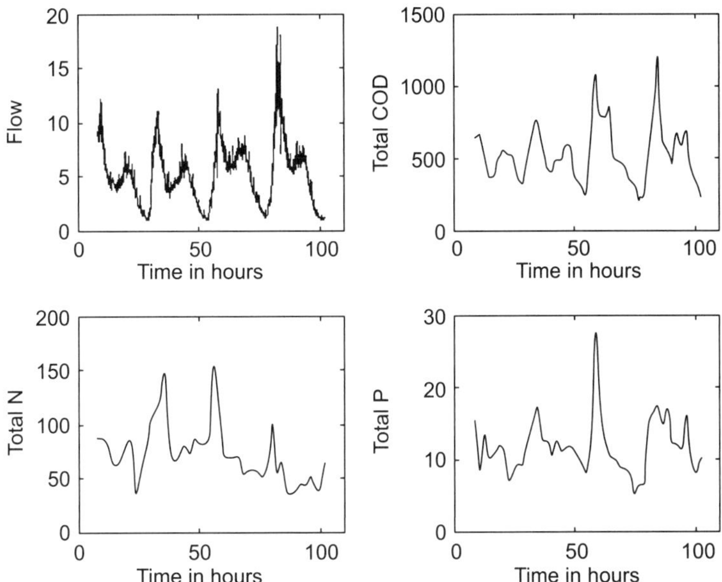

FIGURE 2.8 Typical dry weather diurnal variations in a municipality with mostly household wastewater. The data show variations from Thursday through Sunday. (Note the phosphorus peak on Saturday.)

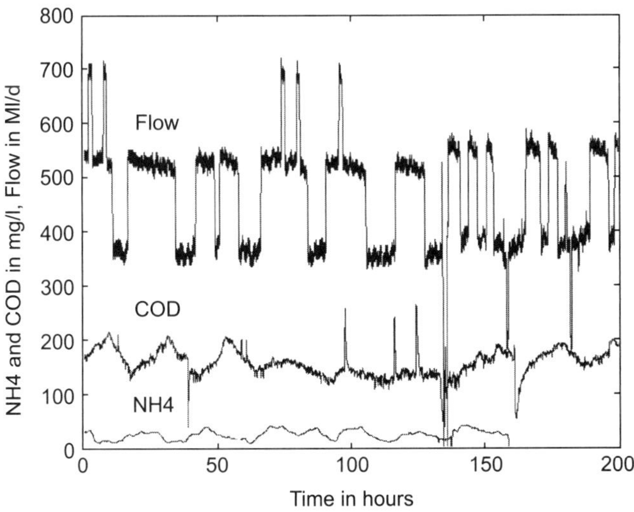

FIGURE 2.9 Influent variations in a large WWTP having only on/off primary pumps, resulting in undesirable sudden flow variations to the plant.

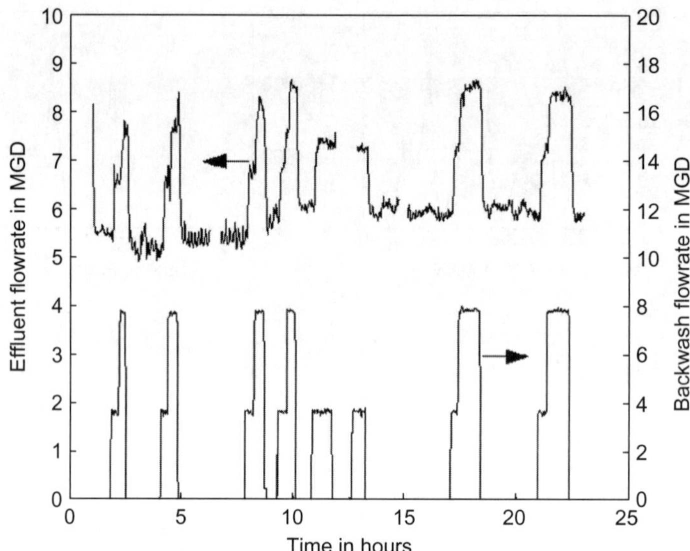

FIGURE 2.10 Filter backwashing (lower curve) and its impact on plant influent flowrate (upper curve) and plant operation.

Disturbances caused by poor operation can be illustrated by filter backwashing. In the actual plant, backwashing increased the influent flowrate by almost 50%, as illustrated in Figure 2.10. The example in Figure 2.10 had an anaerobic reactor as a first step for biological phosphorus removal. The reactor was hit not only by a large flowrate, but also by oxygen-rich water. The water propagated into the next anoxic zone, still with some oxygen left. Obviously, the biological reactions suffered a lot and the effluent quality was unsatisfactory. Apparently, the pumping had to be smoothed. Once the disturbance was understood then the problem was readily solved.

High flowrates will have a significant effect on clarifier performance (Figure 2.11). In particular, sudden flowrate changes caused by a rain storm will create problems in the settler and clarifier, not only as a result of the flowrate amplitude but also because of the high rate of change. Figure 2.11 demonstrates that the clarifier is running close to its maximum capacity and fails during large hydraulic peaks, resulting in large effluent suspended solids concentration values.

If sludge supernatant is recycled to the plant influent during a high load, then the nitrogen load to the plant may be large, as depicted in Figure 2.12. The figure shows

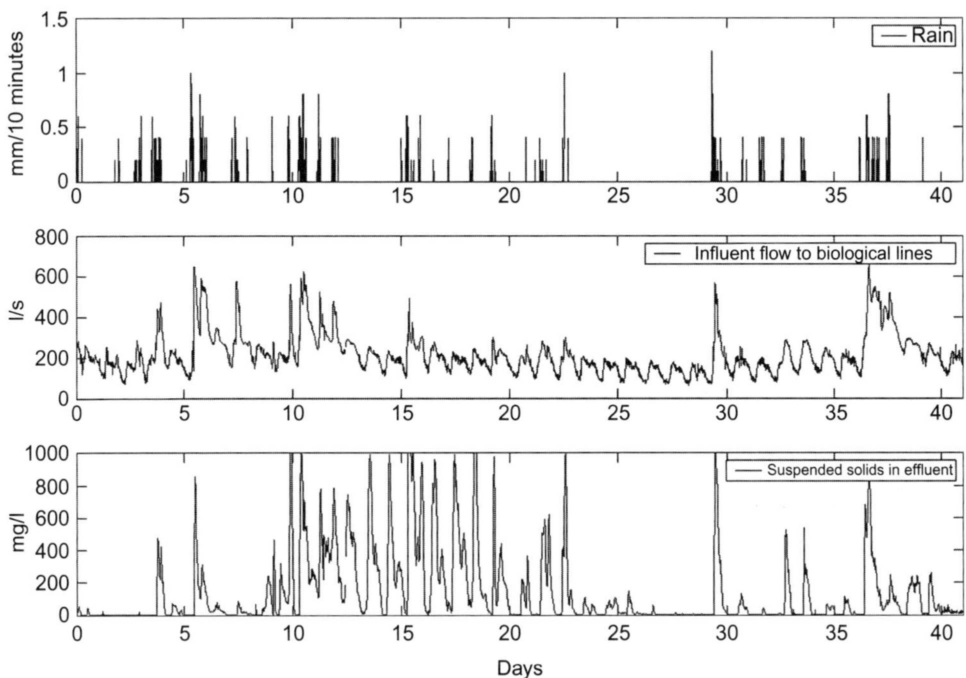

FIGURE 2.11 The relationship between large hydraulic disturbances and effluent quality. The upper curve shows rain intensity during approximately 40 days and the middle curve represents the corresponding influent flowrate to the municipal treatment plant. The lower curve shows suspended solids after the secondary settler.

how the oxygen uptake rate increases significantly as the supernatant is recycled within the plant.

Disturbances also arise from the shift of bacterial populations and the change of their microbial and physical properties. For example, it is not uncommon that a treatment system suffers from sludge settleability problems due to an outbreak of filamentous bacteria. This is often caused by insufficient aeration or food supply to the organisms.

An important part of a control system is monitoring to continuously look for any deviations in the behavior of one or several process variables. Early warning systems are particularly useful in slow biological processes so that disturbances and process changes can be detected before the point of no return.

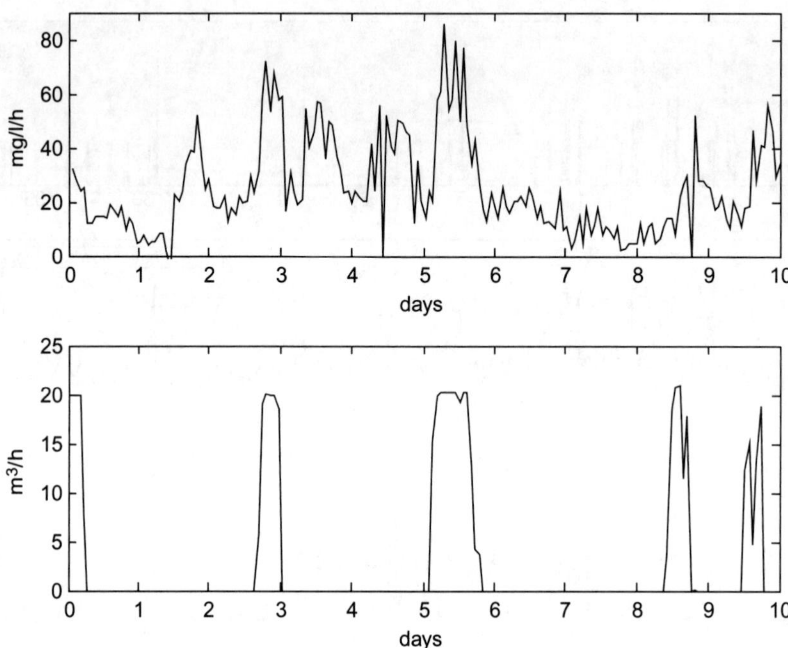

FIGURE 2.12 The effect of supernatant recycling in a plant during a 10-day period. The lower curve shows the supernatant flowrate (which is not very high but has a large concentration). The upper curve shows the oxygen uptake rate in the aerator (Nielsen, 2005).

2.3.4.7 Priorities in Operation

Development toward process/plant-wide control approaches is still in its infancy. The implementations are gaining momentum, but at a low speed.

Often, customer knowledge will limit the use of SCADA systems. To make the systems work requires a proper understanding of the potential of control. Supervisory control and data acquisition manufacturers often do not have an understanding of process models so the controller tunings, the sampling times, or the control structure (i.e., having the correct instrumentation connected to the correct actuator) has to be known. This means that there is a great opportunity not only to use ICA, but to save resources and to obtain consistent operation by better use of the potential in ICA.

Any plant operator must set priorities for proper operation. It is quite apparent that good operation must rely on functioning equipment. All the links in the chain have to be working to get a good operational system. Hardware includes not only instrumentation, but also all the various actuators such as compressors, pumps, motors, and valves. Communication systems are getting increasingly important in plant control systems. The software relies not only on proper control algorithms, but also databases, communication systems, data acquisition systems, human-friendly displays, and, most importantly, people. No control system can be presented to operators who have not been able to influence the design of it; it is all built on trust. Any well-intended and functioning control system can be a total failure if the people operating it do not trust it. Therefore, involvement of the people and education are crucial parts of a successful system. So what are the priorities?

- *Keep the plant running.* Make sure that the equipment is functioning; that the pumps, valves, and motors are operating; that the instruments are calibrated and maintained; and that the signals are properly communicated to the control system. This also includes "low-level control," such as the control of local flowrates, levels, air pressures, or various concentrations that are not immediately connected to effluent quality. Most of these control actions are traditional process control loops, such as air pressure control, liquid-level control, and flowrate control.
- *Satisfy effluent requirements.* It is not sufficient to keep the physical parameters correct. Other variables that are directly related to effluent quality have to be controlled. This is realized at this level. It involves manipulating variables of different unit processes, such as dosage control for chemical precipitation, dissolved oxygen control for aerobic processes, return sludge control, or sludge retention time control. Typically, each one of these control loops represents a simple control loop based on only one process variable.
- *Minimize the cost.* In each one of the unit processes, the control scheme may be more elaborate. One example is dissolved oxygen control, where the dissolved oxygen setpoint should be varying depending on the load. The ultimate goal at this level is to optimize the unit process operation. All of this depends on suitable sensors and instruments. Cost can be influenced by decreasing energy demand (for aeration or for mixing) and by lowering the cost for dosage chemicals in phosphorus precipitation or in centrifuge operation. Cost is also related to personnel. Today, many plants are satisfactorily operated unmanned during evenings, nights, and weekends.
- *Integrate the plant operation.* The purpose of integration is to satisfy the effluent requirement at minimum cost. By coordinating several processes, it is possible

to decrease the impact of disturbances to the plant. The combined operation of processes may make it possible to optimally use the available volumes and the sludge for the best operation.

- Present standard computer hardware and software and the increasing availability of reliable real-time measurements (properly validated) for an increasing range of different parameters enable advanced closed-loop process control on WWTPs, resulting in increased safe operations and better operational economy. However, these benefits can be limited by the design of the WWTPs themselves because the design has not been made with controllability in mind.

2.3.4.8 Instrumentation and Monitoring

Instrumentation (including sensors, analyzers, and other measuring instruments) has developed significantly during the last decade. An increased confidence in instrumentation is now driven by the fact that clear definitions of performance characteristics and standardized tests for instrumentation have become available (ISO, 2003). Commonly used online measurements are summarized as follows:

- Ammonium;
- Biogas production;
- Conductivity;
- Dissolved oxygen;
- Flowrate;
- Level, pressure;
- Nitrate;
- Organic matter;
- pH;
- Phosphate;
- Redox;
- Sludge blanket level;
- Sludge concentration;
- Temperature; and
- Turbidity.

Information needs to be properly extracted from the measured data. Even reliable instrumentation can fail during operation, which can have serious consequences if the instrumentation is used in closed-loop control. Therefore, real-time data validation is needed before using measurements for control purposes. Data validation can be performed by simple methods on measurements from a single instrument or as cross validation on measurements from more instruments if any correlation is expected. If confidence in a measurement decreases, it might be possible (on a short-term basis) to use an estimated value, but eventually control must be set to a default scheme until confidence in the measurement has been restored. Tracking the current process operational state via instrumentation is called *monitoring*.

In a sophisticated treatment plant, there is a large data flow from the process. More instrumentation and new instrumentation development will further provide more data. Unlike humans, computers are infinitely attentive and can detect abnormal patterns in plant data. The capability of computers to extract patterns (useful information) is rarely used beyond simple graphing. Information technology is not commonly used to encapsulate process knowledge (i.e., knowledge about how the process works and how best to operate it). Process knowledge is typically built up from the experience of operators and engineers; all too often, however, it disappears with them when they leave. If process knowledge can be encapsulated, then not only is it retained but the computer can also assist decision making in plant operation. The potential of substantial operator support for diagnosis and for corrective actions is there and has been demonstrated, but it needs to be adopted by the water and wastewater industry.

Most changes in WWTPs are slow when the process is recovering from an "abnormal" to a "normal" state. Early detection and isolation of faults in the biological process are effective because they allow corrective action to be taken well before the situation becomes unfavorable. Some changes are not obvious and may gradually grow until they become a serious operational problem.

Figure 2.13 illustrates daily variations of influent flowrate during approximately 3 weeks. Some significant peaks of the flowrate are obvious. Deviations larger than 3 ought to be observed carefully, and suitable operations have to be implemented.

Figure 2.14 illustrates what happens at a sensor failure. The real measurement is complemented with a high-pass filtered signal. The latter serves as a measure of the variability of the real measurement and can more easily detect any change in the signal character. The filtered signal in the figure demonstrates a significant change in the noise character of the signal, thus revealing a sensor problem.

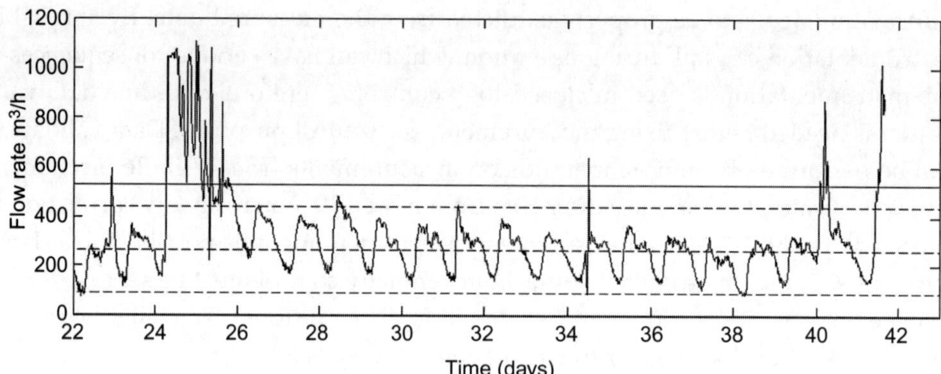

FIGURE 2.13 Influent flowrate variations during a 3-week period. The mean value and the ±2α and ±3α limits are indicated.

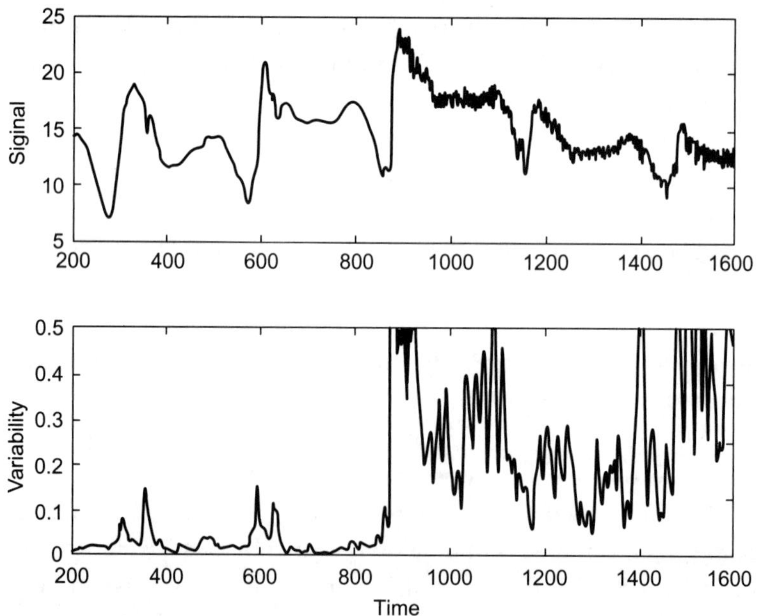

FIGURE 2.14 Detection of a sensor problem. The upper curve shows the sensor signal. The noise character will change after time 900, indicating a sensor problem. The lower curve shows the variability of the signal. When the variability exceeds a threshold (e.g., 0.15), the monitoring system can sound an automatic alarm.

A comprehensive description of control in wastewater treatment systems is available in *Wastewater Treatment Systems. Modeling, Diagnosis and Control* (Olsson and Newell, 1999). An updated state-of-the-art description of control issues in wastewater systems is found in *Instrumentation, Control and Automation in Wastewater Treatment Systems* (Olsson et al., 2005). Sewer operations have been excluded intentionally and the focus is on activated sludge systems.

2.3.5 Laboratory Information Management Systems

A LIMS is a type of software that supports the recording, storing, analyzing, and reporting of the results of laboratory analyses. It provides information about analytical samples received and tested within the laboratory operation. A LIMS can provide information regarding analytical results, status of testing in progress, sample collection data, workload, summary reports and trend analyses for sample analytical results and business operations, as well as quality control information.

The amount of data that must be captured, manipulated, stored, retrieved, plotted, and so forth in a laboratory operation is enormous. Management of these data constitutes a major task of the laboratory operation, second in scope only to the actual processing and testing of samples. A database is simply a collection of such data gathered in a logical format and connected by specified relationships. Laboratory information management systems provide structure for the organized input of data and a means for searching and updating the database to extract information. Effective automation of data management in a LIMS can result in substantial productivity gains and improved organization, control, and accuracy.

Business processes supported by a LIMS follow the life cycle of a sample, as depicted in Figure 2.15. The basic functionality of a LIMS that is relevant to all utilities includes sample registration, data input, and sample reporting. There are potential points of LIMS use at all of the stages.

More extensive functions that are useful to a utility include sample collection scheduling, workload management, and instrument integration for data collection. Scheduling and workload management functionality are useful in assigning and prioritizing sampling and analysis activities and providing notification to staff of pending and overdue tasks. More advanced LIMS functionality includes data capture through bar-code readers for sample tracking and automatic or data-event-triggered output and reporting. This can reduce the workload on staff for data recording in the LIMS and, more importantly, ensure high data quality by removing the need for human translation of data from system to system.

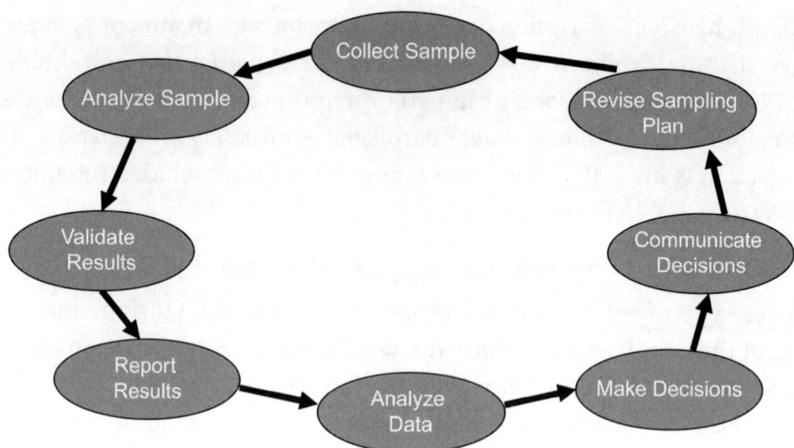

Figure 2.15 Laboratory sample life cycle.

The primary end users of the LIMS are the utility's laboratory technicians or plant operations staff responsible for sampling. They are designed to replace the technician's logbook. In practice, staff may enter the data directly to the system or transfer data from logbooks to the LIMS. Additional stakeholders are users of outputs from the LIMS. This includes plant operations staff responsible for adjusting operating parameters to address water quality, plant management as part of benchmarking and monitoring performance, regulators through the data reporting from the LIMS on standard reports such as Clean Water Act National Pollutant Discharge Elimination System reports, and customers in water quality reports.

The structure of a LIMS is focused on sample data. Data input begins with a sample registration or a "log in," which includes information about the sample, including the person conducting sampling, location of sampling, time and date, and potential climate conditions. The system is then used to define test specifications (i.e., methods, units, and control limits) and to record the results of the analyses. Data entry to the LIMS can be manual or automated through integration of laboratory instrumentation. A LIMS can be configured to import data from standardized external sources such as spreadsheets.

Data quality is an important component of LIMS functionality. Most COTS LIMS software packages include some level of data control and validation on data entry to ensure that data outliers are checked and invalid data are prevented. Methodologies for data quality control vary by system, but may include color coding of fields, error messages, or the inability to save data. Another aspect that contributes to data quality is instrument calibration. Laboratory information management systems also have log capabilities to support calibration recording to ensure the integrity of the data.

Water quality reporting is a regular part of a utility's business requirements. Some of this reporting is needed for internal operations and performance tracking. A utility operation's staff is aided by frequent feedback on treatment processes to support efficient control of the plant environment. Broader analyses of operations can also be conducted through the LIMS to do trend charts and correlation analyses. For example, some systems may include functionality to do dynamic data trending and provide alarms or notifications when there are outliers or changes in trends from the norm.

Regulatory reporting is another important business process supported by LIMS. Many COTS packages have standard reports for the United States that perform the data analysis and calculations necessary and map the data directly to a report that can be sent to the regulatory agency. By having these data available in the LIMS, it is also easier to retrieve copies of reports that have been sent to regulators in the event of an audit or inspection.

Many LIMS also allow for the configuration of calculations defined by end users to support their specific data analysis needs. Data for both internal and external use is typically available for export as electronic files or paper reports.

A LIMS can productively exist as a stand-alone system to support the analytical work of a utility. However, there are opportunities to broaden the use of LIMS data by making final analysis data available to other utility systems. Common areas of integration with other systems are noted in Table 2.7.

Laboratory information management systems are important to utility operations for a number of reasons. They ease the burden of compliance reporting, enable the integration of laboratory test results with plant operational data to improve decision making, and they support the standardization of procedures and ensure repeatability. These lead to better safeguards for data, which are required for professional certifications and requirements for laboratories such as the National Environmental Laboratory Accreditation Conference and U.S. EPA's Good Automated Laboratory Processes (EPA Directive 2185: http://www.epa.gov/irmpoli8/archived/irm_galp/).

As discussed, data maintenance is central to the regular use of the LIMS. Software maintenance for a LIMS is similar to that of any software product at a utility, except

TABLE 2.7 Points of system integration for a LIMS.

System	Integration areas	Purpose of integration
SCADA	Control of sampling devices	Automated sampling and analysis
Maintenance management (CMMS)	Work management	Scheduling of sampling and reporting activities.
GIS	Sample locations	Support analysis of geographic trends in samples Support planning/scheduling of sampling activities

that there may be more extensive software interfaces between the LIMS and laboratory instruments and devices than with other utility information systems. This will require that someone serve in the role of maintaining these interfaces as software is upgraded and/or instruments are replaced.

3.0 REFERENCES

American Water Works Association Research Foundation (2005) *Effective Practices to Select, Acquire and Implement a Utility CIS;* American Water Works Association Research Foundation: Denver, Colorado; p 8.

American Water Works Association Research Foundation (2006) *Asset Management Planning and Reporting Options for Water Utilities;* American Water Works Association Research Foundation: Denver, Colorado.

American Water Works Association Research Foundation (2008) *Asset Management Research Needs Roadmap;* American Water Works Association Research Foundation: Denver, Colorado.

Baumert, J.; Bloodgood, L. (2004) Private Sector Participation in the Water and Wastewater Services Industry's Office of Industries; Working Paper No. ID-08; U.S. International Trade Commission: Washington, D.C.

Box, G.; Jenkins, G. (1970) *Time Series Analysis: Forecasting and Control;* Holden-Day: San Francisco, California.

Boyer, S. A. (2004) *SCADA, Supervisory Control and Data Acquisition*, 3rd ed.; ISA Press: Research Triangle Park, North Carolina.

Clemson University (1987) Simulation of Single-Sludge Processes for Carbon Oxidation, Nitrification, & Dentrification. http://www.clemson.edu/ces/departments/eees/outreach/sssp.html (accessed Aug 2009).

Griborio, A. (2004) Secondary Clarifier Modeling: A Multi-Process Approach, Ph.D. Thesis; University of New Orleans: New Orleans, Louisiana.

Griggs, Neil S. (1996) *Water Resources Management: Principles, Regulations, and Cases;* McGraw-Hill: New York.

International Society of Automation (2003) *Automation, Systems and Instrumentation Dictionary,* 4th ed.; ISA Press: Research Triangle Park, North Carolina.

International Organization for Standardization (2003) *Water Quality—On-Line Sensors/Analysing Equipment for Water—Specifications and Performance Tests,* 1st ed.; 2003–10-15, ISO15839:2003; International Organization for Standardization: Geneva, Switzerland.

International Water Association Task Group on Mathematical Modeling for Design and Operation of Biological Wastewater Treatment (2000) *Activated Sludge Models;* IWA Publishing: London, U.K.

Kavanagh, S. C.; Miranda, R. A. (2005) *Technologies for Government Transformation: ERP Systems and Beyond;* Government Finance Officers Association: Chicago, Illinois.

McCorquodale, J.; La Motta, E.; Griborio, A.; Homes, D.; Georgiou, I. (2004) *Development of Software for Modeling Activated Sludge Clarifier Systems, A Technology Transfer Report;* Department of Civil and Environmental Engineering, University of New Orleans, New Orleans, Louisiana.

Metropolis, N.; Ulam, S. (1949) The Monte Carlo Method. *J. Am. Stat. Assoc.,* **44,** 335–341.

Olsson, G.; Newell, B. (1999) *Wastewater Treatment Systems. Modeling, Diagnosis and Control;* IWA Publishing: London, U.K.

Olsson, G.; Nielsen, M. K.; Yuan, Z.; Lynggaard-Jensen, A.; Steyer, J. P. (2005) *Instrumentation, Control and Automation in Wastewater Treatment Systems.* Scientific and Technical Report No.15; IWA Publishing: London, U.K.

Ramon, G.; Stern, C. (2004) Lessons Learned in Implementing Mobile Computing Technology to Improve Utility Field Operations. *Proceedings of the 2004 Water*

Environment Federation/American Water Works Association Joint Management Conference; Phoenix, Arizona; Mar 14–17; Water Environment Federation: Alexandria, Virginia.

Ramos, L.; Orlov, L. M.; Teubner, C. (2004) *Making Portals Work;* Forrester Research, Inc.: Cambridge, Massachusetts.

Rosen, C.; Ingildsen, P.; Guildal, T.; Nielsen, M. K.; Munk-Nielsen, T.; Jacobsen, B. N.; Thomsen, H. (2005) Introducing Biological Phosphorous Removal in an Alternating Plant by Means of Control—A Full Scale Study. *Water Sci. Technol.*, **53** (4–5), 133–141.

Segal, G. F. (2003) Georgia Public Policy Foundation, Issue Analysis: The Atlanta Water Privatization: What Can We Learn? http://www.gppf.org/article.asp?RT=20&p=pub/Water/atlanta_water.htm (accessed Aug 2009).

Twigg, R.; Payne, J.; Tzaban, Y. (2007) IBM Information Management Software, ECM Taxonomy Management. ftp://ftp.software.ibm.com/software/data/cmgr/pdf/omnifind-DE-taxonomy.pdf (accessed Mar 2010).

Vitasovic, Z.; Zhou, S. P.; McCorquodale, J. A. (1997) Secondary Clarifier Analysis Using Data from the CRTC Protocol, *Water Environ. Res.*, **69,** 999.

Water Environment Federation (2005) *GIS Implementation for Water and Wastewater Treatment Facilities;* Manual of Practice No. 26; McGraw-Hill: New York.

4.0 SUGGESTED READINGS

American Water Works Association (AWWA) Research Foundation (2005) *Effective Practices to Select, Acquire and Implement a Utility CIS*; American Water Works Association Research Foundation: Denver, Colorado; p 8.

Bolton, W. (2009) Programmable Logic Controllers, 5th ed.; Newnes: Oxford, U.K.

Boyer, S. A. (2004) *SCADA, Supervisory Control and Data Acquisition*, 3rd ed.; ISA Press: Research Triangle Park, North Carolina.

Henze, M.; Gujer, W.; Mino, T.; van Loosdrecht, M. C. M. (2000) *Activated Sludge Models: ASM1, ASM2, ASM2d and ASM3*; Scientific and Technical Report No. 9; IWA Publishing: London, U.K.

Henze, M.; van Loosdrecht, M.; Ekama, G.; Brdjanovic, D., Eds. (2008) Biological Wastewater Treatment: Principles, Modeling and Design (UNESCO); IWA Publishing: London, U.K.

John, K. H.; Tiegelkamp, M. (2001) PLC Programming with IEC 61131–3: Concepts and Programming Languages, Requirements for Programming Systems, Decision-Making Tools; Springer-Verlag: Berlin, Germany.

Kavanagh, S. C.; Miranda, R. A. (2005) *Technologies for Government Transformation: ERP Systems and Beyond*; Government Finance Officers Association: Chicago, Illinois.

Chapter 3

Information Technology Planning

(continued)

(continued)

1.0 THE CONTEXT OF UTILITY STRATEGIC PLANNING

1.1 Utility Strategic Planning

The Government Finance Officers Association (GFOA) (Chicago, Illinois) recommends that all governmental entities use some form of strategic planning to provide a long-term perspective and establish logical links between authorized spending and broad organizational goals. In *Recommended Budget Practice on the Establishment of Strategic Plans* (www.gfoa.org/downloads/budgetStrategicPlanning.pdf), GFOA defines strategic planning as "...a comprehensive and systematic management tool designed to help organizations assess the current environment, anticipate and respond appropriately to changes in the environment, envision the future, increase effectiveness, develop commitment to the organization's mission and achieve consensus on strategies and objectives for achieving that mission."

The strategic planning process incorporates perspectives of multiple individuals and groups to ensure the planning process includes landscape element analysis, organization element analysis, and adaptive planning. Landscape element analysis seeks to identify external trends and drivers, such as major businesses entering or leaving the service area, environmental changes in resource supply or quality, changing customer expectations, and looming regulatory requirements. Organization element analysis focuses on internal capabilities and resources, including existing strengths and weaknesses, upcoming changes, and emerging opportunities. Adaptive planning categorizes each of these changes as orderly, dynamic, or chaotic.

A strategic plan typically identifies a limited number of high-level strategies. An action plan defines how these strategies, or goals, will be implemented. Each strategy will have one or more specific, measurable objectives. The strategic plan will also identify ongoing performance measures, sometimes in the form of a balanced scorecard, to help the organization track performance over time.

1.2 The Role of Information Technology Strategic Planning

Information technology (IT) strategic planning is best performed within the context of an organizational strategic plan. In this scenario, IT investments can be linked to organizational goals, objectives, and measures. A certain level of strategic analysis is vital within the IT planning process to ensure IT activities anticipate, leverage, and support changes affecting the utility in a way that best supports the utility's long-term success.

In general, IT strategic plans will identify a series of programs, projects, and action items (also known as "quick wins") for implementation over the period of the plan. Many of the activities used in creation of a strategic plan are repeated within the context of each program and project. The difference lies both in the breadth and depth of application. Strategy covers the entire organization, but at a very high level. Programs cover a specific set of projects and initiatives that are related in some manner. For example, an asset management program might involve implementation of a work management system, data collection activities, and interfaces between various systems. Each of these might be structured as a separate project. In this manner, the depth of investigation increases as the breadth of the investigation decreases. Chapter 4 discusses the processes involved in program management, and Chapter 5 drills down further into the processes involved in project management.

1.3 Overview of the Information Technology Strategic Planning Process

An IT strategic plan is best viewed as a project, with a specific starting point (typically a project kickoff meeting) and a specific outcome (i.e., the published plan). This chapter presents the participants, methods, techniques, and deliverables commonly used in the creation of a strategic IT master plan. Business objectives of the plan are identified and an overview is provided for the plan creation process. Table 3.1 provides

TABLE 3.1 Project planning overview.*

Process	Tools	I/W/S	Participants	Deliverables
Utility strategic planning	SWOT vision	I/W/S	EST	Drivers and objectives
Initiation and kickoff	Vision	W	EST/PT	
Identify business drivers	SWOT	I/W	EST/PT	Drivers and objectives
Review current situation – Business process mapping – Skills mapping		I/W/S	PT/IT/SMEs	
Identification of gaps – Disaster recovery/business – Continuity – Service catalog/SLAs	Gap analysis	W I/W I/W/S	PT IT PT/IT/SMEs	Gaps
Alternatives analysis	Alignment CSFs and KPIs	W	EST/PT	Prioritized opportunities, strategic direction
Plan presentation		W	EST/PT	Program definition, master schedule, and budget estimates

*I = interviews; W = workshops; S = surveys; EST = executive steering team; PT = planning team; IT = information technology employee group; SMEs = subject matter experts; CSFs = critical success factors; and KPIs = key performance indicators.

an overview of the processes, tools, participants, and deliverables discussed in this chapter. In addition, the case study presented in Section 1.0 in Chapter 10 provides an example of one utility's path through this process.

2.0 BUSINESS OBJECTIVES FOR INFORMATION TECHNOLOGY STRATEGIC PLANNING

The primary goal of an IT strategic plan is to ensure IT investments are applied in a way that makes business sense across the entire organization. Key business objectives for performing an IT strategic plan include

- Operational and capital improvement project (CIP) budgeting,
- Business-IT alignment,
- Information technology program development and execution,
- Sustained executive commitment, and
- Chief information officer (CIO) empowerment.

2.1 Operational and Capital Improvement Project Budgeting

The primary business objective of the strategic information master plan is to provide budgetary numbers that can be included in operational and CIP budgets. It is important to identify budget deadlines at the beginning of the planning process, to understand existing allocated funds, and to leverage any existing project justification documentation. The final master plan should include projects that are aligned with budget cycles.

2.2 Business–Information Technology Alignment

Business–IT alignment considers applying IT in an appropriate and timely way, in harmony with business strategies, goals, and needs. The strategic IT plan should focus on alignment around and between

- Business strategies—business scope, distinctive competencies, and governance;
- Organization infrastructure and processes—administrative structure, processes, and skills;
- Information technology strategies—technology scope, systemic competencies, and IT governance; and

- Information technology infrastructure and processes—architecture, processes, and skills.

Table 3.2 provides examples of alignment opportunities.

2.3 Information Technology Program Development and Execution

Information technology projects, across all industries, are especially benefitted by project management techniques. An IT plan packages initiatives into projects that can be budgeted, managed, and implemented discretely using best practices project management techniques. Chapter 4 discusses program development and execution in detail.

2.4 Sustained Executive Commitment

Sustained executive commitment is required to successfully implement IT projects. To achieve this goal, the plan should set forth appropriate expectations of project time, expense, and staffing support requirements. The plan should carefully document the desired benefits to be provided by the project, and each project should be designed in a manner that identifies the business value of each phase. To maintain executive commitment, these business benefits should be tracked and communicated when achieved.

2.5 Chief Information Officer Empowerment

The CIO in a utility may be equal to the directors of finance and operations or he or she may be a manager, frequently under the authority of finance or administration. In some organizations, technology services are provided by a centralized

Table 3.2 Business–IT alignment.

Project	Business strategy	Organization infrastructure and processes	IT strategies	IT infrastructure and processes
Permit tracking system	Allow developers Web access to permit status	Ensure automated or annotated updates of permit status across departments (engineering, finance, Customer Service)	Enterprise system or best of breed? Java or .NET? Single sign-on?	Ensure adequate bandwidth/ security to facilitate access by authorized developers and all participating departments, including field staff, if appropriate

municipal technology group outside of the utility division. As IT project benefits cross departmental (and budget) boundaries, the need increases for empowerment of the CIO, regardless of where the position is located within the organization. This helps ensure new technologies are implemented, configured, and used in a manner that is most beneficial to the organization, not only the department with the original need. Security also demands a stronger collaboration between the IT group and treatment, with an ever-increasing understanding of the risks and concerns of both sides.

A key challenge for a utility CIO is the ability to understand and prioritize between finance/customer service and treatment. The immediate demands of customers who cannot find out information about their account have long been the top priority of a utility's IT department. Treatment, alternatively, uses highly redundant hardware and configurations and, frequently, their own technology staff to ensure zero downtime. If a backup unit fails, it may seem like a lower priority event to the IT group. New organizational structures, which are part of the topic of "governance," require adequate staffing and coverage to ensure the "urgent" does not override the "vital." Service-level agreements, discussed in Section 6.2 and in the case study in Section 1.0 in Chapter 10, can help IT staff understand both the flexibility and lack of flexibility appropriate for supporting treatment.

The strategic IT master plan is a key document to empower the CIO throughout the organization, both ensuring that the CIO understands organizational priorities and helping to communicate those priorities. This high-level document helps each department understand when and how they fit in future plans.

3.0 STRATEGIC INFORMATION TECHNOLOGY PLAN PARTICIPANTS

Strategic plans are typically carried out using two small ongoing groups to provide direction; these are the executive steering team and the project team. A wide variety of individuals in more limited engagements provide depth and breadth to the planning effort.

3.1 Executive Steering Team

3.1.1 Membership and Representation

A steering team is critical to the success of an IT plan. The steering team should include the plan project manager, who is frequently the head of IT. In a small- to medium-sized organization, the steering team should include three to five members of senior management. In a larger organization, a steering team of 8 to 10 mid- to upper-level managers may be more appropriate.

3.1.2 *Roles and Decision Making*

The executive steering team establishes the high-level vision for IT within the organization and for the IT strategic plan, stays abreast of major discoveries during the planning project, and selects alternatives for the projects proposed. The steering team must be able to provide executive insight and direction during development of the IT plan. In small-to-medium-sized organizations, the steering team may be able to officially endorse the plan and submit it to the governing body for final approval. In larger organizations, steering team approval may be a prerequisite to upper management review. As a result, individuals with a good understanding of the process and politics needed for plan approval and success should be sought after.

3.2 Planning Team

3.2.1 *Membership and Representation*

The planning team includes supervisors and power users from throughout the organization who can grasp the potential of the projects proposed, not only for their own department but also for other departments. The IT group should be represented, but should not dominate the membership of this group.

3.2.2 *Roles and Decision Making*

The planning team will review detailed deliverables and participate in detailed analysis of alternatives and prioritization portions of the project. The team will help carry out information-gathering activities and will communicate plan progress to their respective departments.

3.3 Information Technology Group

Input from the IT group is vital, but should be included via interviews, limited participation in appropriate workshops, and one to two representative members on the planning team. This approach ensures appropriate technical depth in a business-focused plan.

3.4 Subject Matter Experts

Subject-matter experts provide knowledge about specific areas, including applications, regulations, and business processes. Their input can be captured using interviews, workshops, and surveys.

4.0 STRATEGIC INFORMATION TECHNOLOGY PLAN PROCESSES

Although there are a variety of ways to carry out a strategic IT plan, typical components that are addressed here include a plan initiation and kickoff task, identification of business drivers, current situation review, identification of gaps and alternatives analysis, and final plan presentation.

4.1 Task 1: Initiation and Kickoff

Project initiation begins with the selection of a project manager for the planning project. Many organizations contract with a consultant to guide them through the planning process. The project manager and/or consultant will create a project schedule and work with management to populate the executive steering team and project team.

4.1.1 Resources and Desk Audit

The following resources are vital to the planning process and should be consolidated for easy access by the project team:

- Annual operations budget;
- Most recent capital improvement plan;
- Most recent organizational chart, updated to reflect current organization;
- Most recent IT asset inventory, updated to reflect current assets;
- Most recent IT configuration diagrams, updated to reflect current implementation; and
- Most recent system maintenance costs, including hardware and software maintenance and fees for recurring services such as leased data or fiber.

4.1.2 Initial Workshops

The following workshops are typically included in the initiation and kickoff phase:

- Kickoff workshop to explain the process, timetable, and team involvement for a project and
- Vision workshop to establish the vision for the plan (i.e., what the plan is supposed to achieve).

4.2 Task 2: Identify Business Drivers

Strategic planning is based on the assumption that change has occurred and will continue to occur. For each change category discussed in the following sections, examples are given of the types of changes that should be considered and the individuals or groups most likely to be aware of changes within that category. Business drivers are typically captured using interviews and workshops.

4.2.1 *Business and Organizational Changes*

Business changes include discrete events such as the appointment of a new top executive and ongoing events such as changing customer expectations. Sample sources of business change include management styles, economic conditions, political climate, and workforce issues. Some changes may not occur without effort, such as potential collaboration partner opportunities. Executives and managers are typically the best source of information regarding business changes.

4.2.2 *Regulatory Mandates*

Regulatory changes can occur at the local, regional (water), state, and federal level. Most utilities also watch California specifically, as many new regulations have been initiated in that state. Capacity, management, operation, and maintenance, with resulting asset management initiatives, are examples of a regulatory requirement with significant IT effect. Consider political, environmental, ecological, and water source issues under this category. Managers, the safety compliance officer, and laboratory staff are the prime sources of knowledge in these areas.

4.2.3 *Performance Improvements*

Managers, customers, and employees can all provide input on where performance improvements could provide value to the organization.

4.2.4 *Changing Technologies*

Consider how work processes changed with the advent of the fax machine, supervisory control and data acquisition, and the Internet. What emerging technologies have the potential to significantly change how required processes are performed? How will expectations change when automatic meter reading is widely implemented? Leverage your creative thinkers wherever they are found in your organization. Tap the knowledge of industry experts, regulators, consultants, contractors, vendors, your internal IT staff, employees, and customers. Consider vulnerabilities and potentials, infrastructure, and applications.

4.2.5 *Cyber Security*

Treatment, maintenance, and engineering staff know the locations of physical assets at higher risk for your organization. Your IT staff knows the main access points for technology attacks. Take advantage of this planning exercise to dig deeper and identify unsuspected access points, such as modems, wireless access points, and direct connections to the Internet at remote locations that may have sprung up based on legitimate business needs without the knowledge of the IT staff. Take this opportunity to build a broad view of technology security needs so that the identified solutions are adequately far-reaching. Consider having a security audit performed by an external expert for adequate depth and input on current issues and alternatives.

Industry organizations exist that can provide resources and direction in assessing and ensuring the cyber security of your organization. The American Water Works Association (Denver, Colorado) hosts an annual Water Security Congress, WaterISAC, which issues periodic updates to subscribers; the Department of Homeland Security has also developed a Control Systems Cyber Security Self-Assessment Tool (CS2SAT) that is available to Water Environment Research Foundation (Alexandria, Virginia) subscribers and Water Research Foundation (Denver, Colorado) members.

4.3 Task 3: Review Current Situation

The project team reviews the documents gathered in the desk audit and the information captured in the kickoff and business-driver tasks. Additional investigations are conducted to develop a deep and broad understanding of the current IT situation.

4.3.1 *Identify Key/Relevant Business Processes*

Based on the goals and objectives identified by the steering committee and the tools and methodologies discussed in this report, the planning team should seek to identify business processes with potential for improvement. These processes should be prioritized based on both the vision, goals, and objectives of the IT plan itself, but also the vision, goals, and objectives of the organization as a whole. Evaluating potential improvement opportunities against organizational goals will help the team identify key business processes to be investigated during the planning process.

4.3.2 *Assess Current Conditions*

Selected business processes should be documented in their current state. If possible, capture measurements to quantify the issues, such as the time required to complete one iteration of the process and the number of times per period the process is repeated. The goal is to fully understand the current needs and requirements so that not only the "what" but also the "why" is clear to the planning team.

4.3.3 Identify Desired State (With Timeline)

The next step is to design a better tomorrow. Based on industry-available tools and/or custom applications, what is a best-case scenario for the selected business process? What would be needed to achieve this endpoint? Identify time constraints, including factors preventing change (i.e., not until) and factors forcing change (i.e., must change by). This work transitions into the next task: identification of gaps.

4.4 Task 4: Identification of Gaps

Information collected in current situation review, vision, interviews, and workshops reveals "gaps" between where the organization is today and where the organization desires to be in the future. Most gaps represent opportunities for improvement. Opportunities that can be resolved based on management approval (such as purchase of a new printer in a specific area to meet a specific need) are considered quick wins. Opportunities that require more time and financial resources should be defined as potential projects for further evaluation. It is important to document all gaps, at least briefly, even those that clearly cannot be addressed during the current planning cycle. Changing technologies and other factors may provide unexpected opportunities that might be missed if the need was not documented.

4.5 Task 5: Alternatives Analysis

4.5.1 Identify Alternatives

The purpose of this task is to identify various options for taking advantage of the opportunities identified in the previous step. The "do nothing" alternative should be included in every evaluation and should include an estimated range of life expectancy.

Small opportunities may have one clearly defined option. Larger opportunities merit more analysis and may have three or more alternatives worthy of investigation. For example, customer information system alternatives might include do nothing, minor upgrade with current vendor, major upgrade with current vendor, move to new product with current vendor, and full procurement process for new vendor/application.

Each alternative should have an identified range of budget, time, and support requirements. Ongoing support costs and interim staffing requirements should be identified where appropriate. The goal, at this point, is to understand the issues without selecting a definitive path forward.

4.5.2 Prioritization

Selecting which projects to approve and which alternative for each project should be based on clearly defined and supportable prioritization criteria. All recommended projects should be analyzed using the same criteria, although not all criteria will apply to all projects. Prioritization affects not only what projects are approved, but also the implementation sequence. Prerequisites should inherit the prioritization factors of their dependent projects.

Prioritization factors should include utility strategies, if available, and costs, value, and risks. Factors should be weighted and some factors, such as the ability of the utility to carry out the project, may have a go/no-go effect on the evaluation process. If a prioritization process reveals a ranking that seems wrong to the project team, prioritization criteria should be reevaluated to identify missing factors. The goal is to clearly define and document the actual prioritization process.

4.5.3 Develop the Plan Schedule

4.5.3.1 Organizational Considerations

Staffing effects of projects under consideration should be identified. Effects to IT and department staff in the areas where the new technology will be implemented should be included. Back-fill and temporary staffing potential should be evaluated. The designer should consider using consultants for specialty skills, tasks that are repeated only once every several years (such as this planning process), and in areas where the utility does not have core competency in-house.

4.5.3.2 Budget Considerations

It is important for the designer to understand the availability of funding over the plan period. In addition to understanding the potential annual budget and capital improvement funds available, some funding may be tied to other, non-IT projects, while other funding may be department- and/or project-specific.

4.5.3.3 Political Considerations

Upper management and the public relations officer can provide input on local political considerations that may affect project timing and approval. Upper management and regulation specialists can identify political factors on a local and federal level that may increase the priority of specific projects.

4.5.3.4 Phased Approach (1 Year, 3 Years, 5 Years)

The IT planning process should be repeated every 5 to 10 years, with progress assessment and midterm updates conducted every 1 to 3 years. Once a prioritized list of projects has been developed with an understanding of the staffing, budget, and political factors influencing project timing, each project should be evaluated for project phasing. Some projects will have clear requirements, selection, and implementation phases, while others will proceed straight to implementation.

Start applying projects to your planning calendar by identifying the project with the highest priority and longest timeframe. Fill in the calendar with projects of decreasing priority and project size. Adjust the schedule to level out the effect on the budget and the staff. Fill in with smaller, lower-priority projects during lulls between major project phases.

4.5.3.5 Designing Projects to Obtain Desired State

Chapter 5 describes how to define and conduct IT project plans. Although an IT plan should not contain detailed project plans, an understanding of the work required to complete each project will help ensure an appropriate overall project schedule.

4.6 Task 6: Plan Presentation

The proposed projects with their prioritization criteria should be presented to the steering team for input, review, and approval. Some organizations will consider the plan complete at this point, while other organizations will need to present the final plan to their governing body for approval. The plan should be actionable and approved projects should have clearly defined outcomes.

5.0 METHODOLOGIES FOR INFORMATION TECHNOLOGY STRATEGIC PLANS

5.1 Strengths, Weaknesses, Opportunities, and Threats Analysis

5.1.1 Strengths, Weaknesses, Opportunities, and Threats Definition

The concept of strengths, weaknesses, opportunities, and threats (SWOT) is illustrated using a quadrant as shown in Figure 3.1. Strengths and weaknesses are internal to the organization, and opportunities and threats are external.

5.1.2 How to Conduct a Strengths, Weaknesses, Opportunities, and Threats Analysis

The length of time required for a SWOT analysis is relative to the size of the group, the level of complexity of the external and internal environments, the skills of the people involved in data collection and analysis, and their understanding of the issues being presented. A small group might conduct a SWOT analysis in an hour

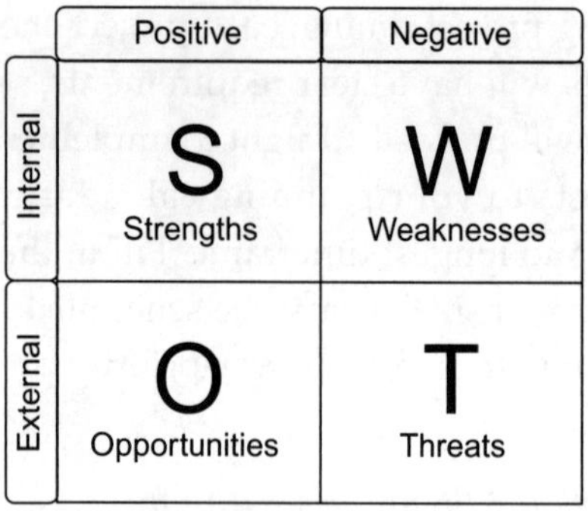

FIGURE 3.1 Strengths, weaknesses, opportunities, and threats analysis quadrant.

or two, while a larger group might spend a half-day on the exercise. The number of people involved is less important than the skill sets of the individuals involved. Consider including upper level management, regulation specialists, public relations, and consultants. Start with the upper left quadrant, strengths, and list the strengths of the organization. Move to weaknesses when no more strengths are forthcoming. Allow additions to all previous quadrants, but postpone suggestions for future quadrants.

5.1.3 *How to Use a Strengths, Weaknesses, Opportunities, and Threats Analysis for Strategic Planning*

When the quadrants have been populated, evaluate each of the items for their potential effect on IT within the organization. Seek to maximize strengths and opportunities and minimize weaknesses and threats. A SWOT analysis is frequently included as part of the current situation review discussed in Section 4.3 and feeds into gap analysis discussed in Section 4.4.

5.2 Interviews, Workshops, and Surveys

5.2.1 *Interviews*

Interviews provide an extended time period for individuals to provide broad input to the planning process. Interviews with executive management provide insight into orientations toward IT and help identify critical factors for evaluating potential projects. Key customers can provide an outside perspective on critical public-facing processes. Interviews are an effective way to investigate detailed issues raised in other meetings.

Consider having utility managers conduct stakeholder interviews across department boundaries. This approach will broaden managers' perspectives on how data and technology are used in other departments and change the dynamics of the conversation compared to an interview conducted by a superior of a subordinate (i.e., from a problem-solving discussion to a broader discussion of general issues and possibilities).

5.2.2 *Workshops*

Workshops are a proven method of quickly building a broad, robust understanding of a specific set of issues. Workshop organizers should strive to ensure the workshop provides business value commensurate with the time commitment required by all the participants. The organizer and participants should have a clear understanding of the outcome of each workshop. Typical workshops include

- Kickoff workshops;
- Vision workshops;
- Business drivers workshops;
- Process flow mapping workshops for key, cumbersome, or error-prone processes;
- Alternatives evaluation workshops; and
- Prioritization workshops.

5.2.3 *Surveys*

Surveys can be used to ask a limited number of questions of a broad number of individuals. Surveys can help build data to support plan proposals if the questions are multiple choice vs essay. Essay-type questions require significantly more time to complete and should be avoided wherever possible. Weigh the value of each question against the cost to the organization of having all responders answer the question. Ask what difference it will make if the answer to each question is known, and delete questions that do not have the potential to clarify issues, support initiatives, or provide other specific information.

5.3 Alignment with Organizational Vision, Goals, and Initiatives

An IT Plan will be better focused and better received if it is aligned with an existing organizational vision, goals, and objectives. Depending on the approach used, an organization may have some or all of the following in place:

- Vision,
- Mission,

- Goals,
- Objectives,
- Critical success factors, and
- Key performance indicators.

All of the aforementioned elements can be applied at the following levels (each level should support the levels above it):

- Organizational vision/goals/objectives;
- Information technology vision/goals/objectives (department level);
- Information technology plan vision/goals/objectives (plan level); and
- Information technology project vision/goals/objectives (project level).

5.3.1 *Vision*

A vision statement should be short and clear. Most people find it difficult to write a concise, meaningful vision statement. Longer statements are easier, but may not yield the benefits in alignment of employee behavior that are possible when a short, clear vision is well communicated. Frequently, executives are skillful at synthesizing complex issues to short, clear statements that can be beneficial in the vision development process.

To start the development of a vision statement, list words important to the organization in the delivery of IT services. Look for redundancy and cut out words with the least effect. If the team contributes sentences, capture them all, then look within the sentences for key words and phrases. If a short, clear vision statement cannot be obtained within a reasonable amount of time, move ahead with the shortest version that receives approval by the group.

5.3.2 *Goals*

Goals should be actionable and business related. Goals may be as short as a few words and should be no more than two sentences each. Twenty is a reasonable, but not precise number of goals. If the list grew to 50, some consolidation would be appropriate.

5.3.3 *Objectives*

Objectives should be measurable proof that goals have been met. The steering committee will have both broad and specific goals at the beginning of the IT planning project. Working with the steering team to quantify these goals will increase the planning team's understanding of the goals and increase the possibility of the plan achieving these goals.

5.4 Critical Success Factors and Key Performance Indicators

Critical success factors define key areas of performance that are essential for the organization to accomplish its mission. Critical success factors are general in nature and most are industry specific. Key performance indicators are measureable and help management gauge organization effectiveness in support of critical success factors. Critical success factors can be identified at an organizational, departmental, and project level. This discussion focuses on critical success factors at the project level.

5.4.1 *Business Centric*

Factors supporting approval of a project are good launch points for the development of critical success factors for an IT project. Analysis performed during the planning project can help identify key performance indicators that the organization expects to improve as a result of the project. If possible, key performance indicators should be put in place before the project so that a before-and-after perspective can be obtained.

5.4.2 *Executive Involvement*

It is important to identify "measurements that matter" to the organization's management. What will management do based on these numbers? It is wise to consider how supervisors will investigate issues that are brought to the attention of management.

5.4.3 *Focus on Implementation of the Plan*

In addition to factors relating to the business effect of the project, some factors are common across projects, such as whether the project is on time and within budget. Projects should be developed in a manner that best supports early warning of project derailment.

6.0 STRATEGIC INFORMATION TECHNOLOGY PLAN OPTIONAL PROCESSES

6.1 Disaster Recovery and Business Continuity

Disaster recovery and business continuity should be included at a high level in the strategic planning process. Consider where your organization could relocate your servers in the event of a disaster that made your primary server room unavailable for an extended period of time. Consider, also, high-speed connectivity between locations. If this is economically feasible, hot standby could be an option for your organization. Even if not, disaster recovery can be meshed with the need for staging servers to maximize the value of your investment and increasing the likelihood that your disaster recovery equipment will be capable of supporting your organization in the event of an emergency. A full disaster recovery and business continuity plan, which is frequently performed as a separate project, should identify the utility's critical functions and personnel and ensure data, applications, and workstations will be available.

A full disaster recovery plan should include the following steps:

- Identifying and prioritizing IT applications, networks, and other services.
- Which IT assets will need to be recovered right away vs those that recovery could be delayed for a month or more?
- What backup systems could be used (such as non-network-dependent laptops, paper and pencil recording, etc.) if the network is down for an extended period?
- How might staff access critical databases if backup servers are placed 50-plus miles away and both phone lines and automobile roadways are not usable?
- The urgency for this type of planning may be area-specific, but all agencies should have this type of plan, which can be implemented when needed.

6.2 Service Catalogs and Service-Level Agreements

A service catalog, at the most basic level, is a list of services provided by an entity. A service-level agreement (SLA) is an agreement between two parties on the level of services to be provided, communication protocols in case of service interruptions, and escalation procedures in case of inadequate performance by either party. Service-level agreements are intended to increase understanding by the service provider of the consequences of service interruptions and by the customer (i.e., entity receiving service) on the alternatives for reducing or mitigating the consequences of significant service interruptions and the cost implications of those alternatives.

Service-level agreements are an effective tool for use by IT groups in establishing and maintaining priorities during service interruption incidents. At a minimum, SLAs should cover the following topics:

- Duration of agreement;
- Clear identification of parties involved in the agreement, including primary and backup contact information for each party;
- Services covered by the agreement;
- Normal service levels;
- Service levels at which contingency plans and/or penalties start to accrue; and
- Escalation protocols and contact information.

A service catalog is a reasonable precursor to the creation of an SLA. In some organizations, the service catalog serves as a living document, providing consolidated information on available services, primary and backup support providers, and other key information necessary to facilitate problem resolution.

Service-level agreements are an effective tool in establishing business need for resources, including infrastructure and staffing, as they identify business effects of service outages and the costs associated with more reliable service levels.

6.3 Select Business Process Mapping

Business process mapping is an effective tool for identifying opportunities for business process improvement based on implementation of technology tools. Flow charts illustrate processes, decision points, and outputs of the business practice being evaluated. Various methodologies exist. The most common form of flow chart is shown in Figure 3.2. Small circles or ovals represent the start and stop of the process, and are optional. Rectangles represent discrete steps or processes. Diamonds represent decision points. Parallelograms represent input/output. Arrows connect the other elements to show flow through the process. A swim-lane process flow adds rows or columns (at the author's discretion) to add an understanding of the groups or individuals responsible for each step in the process. The flow chart in Figure 3.2 is redrawn as a swim lane in Figure 3.3.

The benefit of a flow chart is that it provides a picture of process complexity, thereby facilitating identification of opportunities for process improvement. Quality experts recommend ongoing mapping of business processes, both for documentation

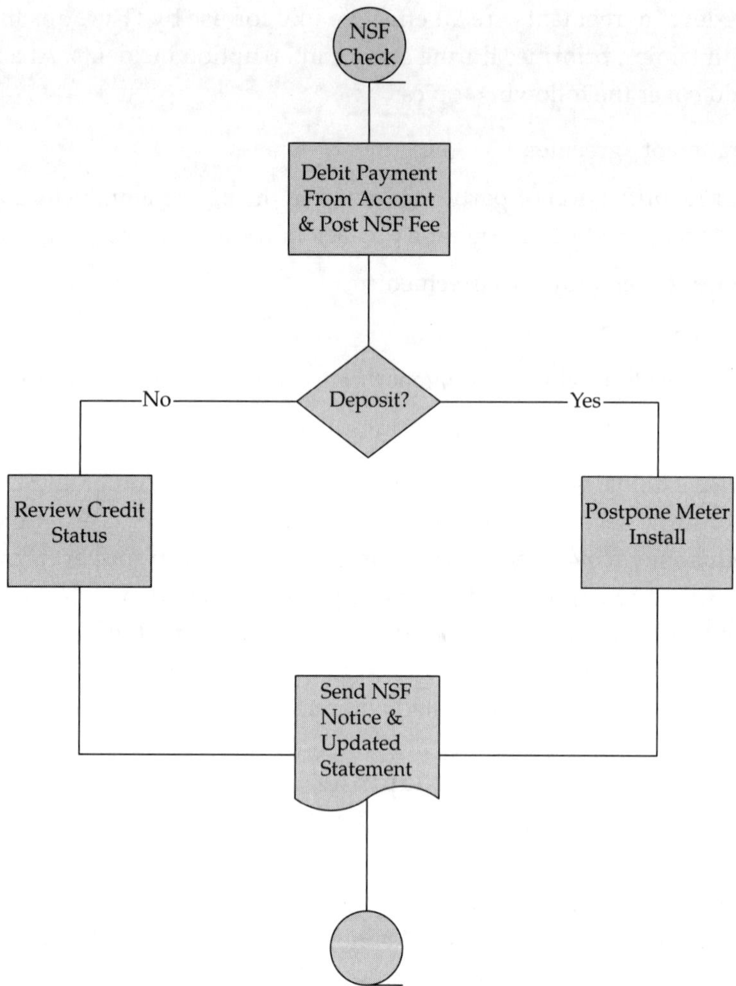

FIGURE 3.2 Business process mapping flowchart (NSF = non-sufficent funds).

on how to perform a task in a standard manner and to facilitate ongoing process improvements. Flow charts can be created to document existing processes and modified to show "to be" processes based on proposed IT investments.

6.4 Skills Mapping

Skills mapping involves identifying the skill sets of existing personnel with current and projected skill requirements. This is especially relevant, for example, if the

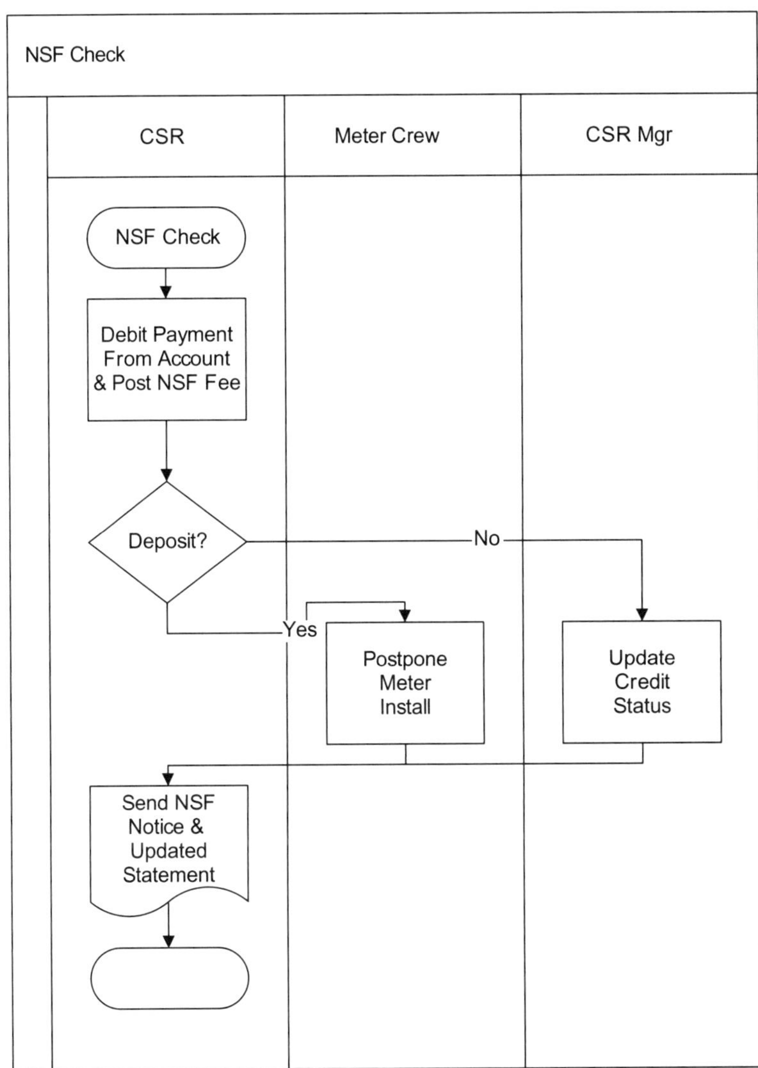

FIGURE 3.3 Swim-lane process mapping flowchart (CSR = customer service representative and NSF = non-sufficent funds).

strategic IT plan proposed moving from a homegrown system to a COTS system based on an industry-standard database. Skills mapping can also be performed on a broader scale to identify overall training needs within the organization. For example, business process improvements are enabled in many organizations by having an appropriate distribution of business users with in-depth understanding of their primary applications. It is also valuable to measure the ability of novice users to have first-level technology questions answered by a nearby business user. In many cases, these questions reveal a lack of understanding of business processes, creating a broader learning opportunity that a remote help desk might miss.

7.0 PRIMARY DELIVERABLES FROM INFORMATION TECHNOLOGY STRATEGIC PLANNING

7.1 Prioritized Drivers and Objectives

The output of the business drivers' workshops should include a prioritized list of drivers and prioritized objectives to address key drivers.

7.2 Identification of Gaps

Gaps should be identified as soon as they are revealed and included as a clearly defined section in each applicable project deliverable, including workshop and interview summaries. For medium-to-large planning efforts, pulling all gaps into a separate document is typically required.

7.3 Prioritized Improvement Opportunities

Pull identified projects into a concise format for easy review by the steering team. Include the project name, factors that affect prioritization, and the resulting prioritization ranking. The list is typically sorted in prioritized order.

7.4 Preferred Strategic Direction

The steering team typically adjusts prioritization factor scores and/or weights to reflect their understanding of how projects support organization strategies. This adjusted prioritized list establishes the preferred strategic direction of IT resources during the range of the plan.

7.5 Program Definition, Master Schedule, and Budget Estimates

Once the plan has been approved, the next steps are to break the approved projects into manageable programs, to develop a high-level program master schedule, and to develop high-level annual program budgets for inclusion in operating and capital improvement budgets. Chapter 4 addresses programs in detail.

8.0 AFTER THE PLAN

In *Recommended Budget Practice on the Establishment of Strategic Plans*, GFOA calls for yearly review and adjustment of the overall IT strategic plan. Current programs, projects, and initiatives should be compared to the original goals to ensure anticipated results remain possible based on current program and project status. In some instances, industry and/or technology changes can dramatically shift priorities, justifying a significant change in direction. In other instances, however, gentle, course adjustments have unexpected consequences that can be remedied if the original goals are reviewed on a periodic basis.

9.0 SUGGESTED READINGS

Collins, J. C. (2001) *Good to Great: Why Some Companies Make the Leap—and Others Don't*; HarperBusiness: New York.

Goodstein, L. D.; Nolan, T. M.; Pfeiffer, J. W. (1993) *Applied Strategic Planning: A Comprehensive Guide—How to Develop a Plan that Really Works*; McGraw-Hill: New York.

Government Finance Officers Association (2005) Recommended Budget Practice on the Establishment of Strategic Plans. http://www.gfoa.org/downloads/budgetStrategicPlanning.pdf (accessed Aug 2009).

Wikipedia, SWOT Analysis Web Page. http://en.wikipedia.org/wiki/SWOT_analysis (accessed Aug 2009).

Chapter 4

Developing an Information Technology Program for a Municipal Agency

(continued)

1.0 THE BASICS OF A UTILITY INFORMATION TECHNOLOGY PROGRAM

1.1 Intent

An information technology (IT) program is a planned and orchestrated assembly of efforts and projects designed to support or enact specific business objectives. An associated program plan is the documentation of these intentions. Such efforts can be ongoing operations and maintenance (O&M) improvements, specific one-time O&M, capital projects, or any combination of these. Information technology programs may come out of a strategic IT master plan, as described in Chapter 3, or may be triggered by some other event or process. The objectives to be met are typically business objectives, not technical objectives. Indeed, as discussed previously in this manual, IT is not the end, but a means by which business processes are performed. The basic intent of an IT program then is to implement IT changes and enhancements to support a customer's business requirements.

Given that a program is designed to support a utility's business, it is important to first understand some basic business variables. For utilities and municipalities, the term *business* is different from conventional use; the core business of a water and/or wastewater agency is to treat water or wastewater. Accordingly, core business functions for this industry include

- Planning,
- Acquisition,
- Conveyance,
- Treatment,
- Distribution,
- Meet regulations,
- Engineer systems,
- Construct systems,
- Operate and control systems, and
- Possibly even buy or sell power.

In more general terms, an assemblage of functions such as the aforementioned is often referred to as *utility business architecture* (UBA). Brueck et al. (1997) summarized UBA as follows: "UBA includes a generic model that identifies each major function within the utility using a 'value-chain' from water resources, to treatment, to distribution, to customer use. UBA also allows a utility to examine and control its various functions and to show the relationships and interactions among these functions." Secondary, or support business functions, may include

- Accounting,
- Billing,

- Finance,
- Human resources, and
- Purchasing.

According to the American Productivity and Quality Center's (Houston, Texas) 2008 Process Classification Framework (http://www.apqc.org/portal/apqc/ksn/PCF_5x.pdf?paf_gear_id=contentgearhome&paf_dm=full&pageselect=contentitem&docid=152203), such support functions have been standardized across industries as follows:

- Develop and manage human capital;
- Manage IT;
- Manage financial resources;
- Acquire, construct, and manage property;
- Manage environmental health and safety;
- Manage external relationships; and
- Manage knowledge, improvement, and change.

These functions are considered support, not because they are less important, but because they are not tied *directly* to meeting core business objectives; a water utility is in business to deliver clean water, not water bills (although the bills are certainly necessary).

Whereas the program may cover all of these business processes, it likely covers a smaller subset, typically defined by organizations such as operations IT or business IT, or can even be limited to strictly finance or control systems. Regardless, it is important to understand the business universe the program will address, not just the technical details.

These core business processes encompass a short list of fundamental elements to consider in an IT program, and include

- People,
- Objectives,
- Practices, and
- Technology.

Fundamentally, then, an IT program is about helping people meet their objectives by using the appropriate tools. It is as much, if not more, about people and business processes than it is about technology. It is not recommended that a program be started with the foregone conclusion that simply installing new hardware or software will meet an objective. Instead, it is important to understand the business objectives people possess, what process and data they need, and then derive the technology solution from these requirements. It may be found that technology is actually the smaller part of the effort, and that, to truly solve the problem, a change in organization or process is needed instead. In the end, all three (i.e., changes in organization, process, and technology) will likely be needed in varying degrees.

This chapter discusses the basis contents for a program plan. More importantly, because every program is different and no "cookbook template" exists, this chapter will also discuss the methods, activities, and pitfalls that will need to be executed or avoided in creating a successful program.

It is important to note that there is a difference between the content of a program plan and the actions that may be needed to develop one. A suggested program-plan document outline is shown in the following section; however, there are several important steps that may need to be taken to deliver the proper content of this outline. The following section discusses content and outline, whereas later sections discuss what actions may be needed.

1.2 Content

The content of an IT program plan can vary widely. Although no strict guideline exists, the following is a generic outline that can be expanding or trimmed, as necessary:

- Approval/signature page;
- Executive summary;
- Discussion of current problems and/or opportunities;
- Discussion of objectives and requirements to be met and why;
- Discussion of any constraints or limitations;
- Discussion of "as-is" state of the system;
- Discussion of "to-be" state of the system;
- Discussion of gap analysis (the difference between as-is and to-be states);

- Discussion of options evaluated and selected to fill the gap;
- Strategic discussion on how to get from here to there (i.e., what needs to happen):
  - Organizational effects,
  - Business process effects, and
  - Technical effects;
- Tactical plan on how to get from here to there (i.e., how it needs to happen):
  - Specific changes to the organization;
  - Specific changes to detailed business processes;
  - Critical success criteria;
  - List of specific projects, with schedule, resource, and budget requirements;
  - Discussion of interrelationship and links between projects; and
  - Summary schedule, resource plan, and budget;
- Risk analysis and mitigation;
- Business case summary;
- Next steps; and
- Conclusion.

The aforementioned outline is relatively simple and self-explanatory. Whereas the specifics may vary from organization to organization, it is a straightforward layout of information that should walk interested parties through from a problem definition to a logical solution and leave readers with a clear sense of understanding and resulting support. The program plan should be easy to read, in layman's terms, and with a business sense to it. Technology, as described in the plan, is presented as a *derived* response to business needs, and should also be presented in layman's terms as well. Conversely, it is not recommended that the program plan be a dissertation on hardware, software, networks, etc.; these details should be handled elsewhere, perhaps as an appendix or separate technical study for those interested in the technology details. Such technical details should, however, be ready as backup that there is a solid foundation to the final recommendations.

1.3 Process

In stark contrast to the aforementioned outline of a program plan, the process by which the program plan is developed, approved, and eventually executed is as unique as each organization and scope itself. There are numerous factors to consider, and they apply in varying degrees to each environment. As such, there is no set process or formula to provide and no template to fill out; guidelines are provided here instead.

A primary program development objective to remember is that, in the end, it is all about supporting people. Indeed, people define the problem and set the objectives, and they need to agree with an assessment of the challenges, options, and solutions. People need to agree with the program plan to a degree that they will fund it and trust it to deliver what was promised. Finally, they need to be happy with the results, for they will be the ones that determine if it is a failure or success. Although focused on logistics of developing a plan, much of this chapter has an interpersonal and qualitative flavor to it; if these human factors are not considered, the plan will likely be challenged.

The following section addresses likely variables in developing an IT program plan. Again, these may represent different issues than those presented in the aforementioned outline.

2.0 UNDERSTANDING PROGRAM PLANNING VARIABLES

This section discusses some of the more qualitative and less methodical aspects of successful IT program planning. By understanding these, the program manager will be better equipped to develop and execute an IT program. Many of these variables, such as consensus building and business process understanding, will not be found in the body of the plan itself; rather, they are factors and activities that will help lead to the plan's development.

2.1 Program Planning Variables

Developing an IT plan for a water and wastewater utility is a unique and often singular effort. Each plan, even if for the same utility but at a different time, will be distinctive. Considering such variables and factors will help determine how and where energy will need to be spent. Understanding these factors will aid in developing the content of the program plan document itself, without which the content of the plan may omit key decision points. Many of these factors have been shown to be critical to program success and/or failure. As an introduction to concepts, these issues include

- Executive support;
- User involvement;
- Clear objectives and requirements;
- Project management;
- Scope control;
- Drivers and constraints (i.e., causes and limitations) and understanding of as-is and to-be states;
- Business processes and business environment knowledge;
- Consensus building, lobbying, and partnering;
- Managing communications, expectations, motivation, and change;
- Business case and justification (both qualitative and quantitative);
- Risk assessment and mitigation;
- Documentation; and
- Packaging and presentation.

Most of the aforementioned factors will be discussed in this chapter; although important, some factors are too specific to each organization to elaborate upon here and, therefore, should be considered locally.

As seen from the aforementioned list, technology is a limited piece in the puzzle. Program planning is not limited to technological skills alone; rather, it is a combination of organizational and logistical disciplines, business, technology, and human factors. Although there are general approaches to planning, if one considers the aforementioned variables there is no single solution to successful program development.

2.2 Executive Support

Executive support is a critical component in producing and executing a program plan. The role of executive support is to advance and advertise the message and objectives about a program, exhibit sponsorship, "rally the troops," procure the resources (staff, time, and money) needed to prepare and execute the plan, and to remove or minimize obstacles along the way. Conversely, a lack of these functions can present significant challenges for the program to overcome.

Executive support is one of the first things a program manager should procure. Such support will aid in further development of the plan as needed resources will be easier to secure. Early executive support may result in scope modifications or refinement to better meet executive expectations and objectives; however, these changes will also result in some level of ownership and commitment. Executive support should also provide an ally for the program manager to turn to.

2.3 User Involvement

Second to executive support, user involvement is a significant target to engage. Users are the eventual true owners of a system and need to help define not only existing obstacles and opportunities, but the vision of the future state as well. Their involvement early on in the process helps garner support for the program, commitment of critical user resources in the plan's development, and helps to shape and prioritize activities and deliverables. In effect, a program manager is a facilitator whose job is to deliver a user's needs.

User involvement assists in defining and clarifying these needs (i.e., requirements) and focusing the associated scope and deliverables. Early on in the process, it is important to identify and engage the user base, or at least a representative group of them, to be an active part of the program process. If possible, it is important to make sure these users will be part of the program team through final acceptance of the system. A few key personnel should be selected to be the final "owners" of the system; these personnel should be made aware of this role from the beginning as they will be the "go-to" people when the program is closed.

2.4 Objectives and Requirements

Objectives, goals, or changes to be implemented must be clearly understood by all involved from the beginning of the program. This is not to say that objectives and requirements are absolutes and can never change, rather, that even the changes must be understood and agreed upon.

Both executives and users play a critical role in defining objectives and requirements. It is important to get objective and requirements clarified and written down from the beginning as they will be the basis of all the work that will be undertaken. These are considered the basis of the strategy, often derived by strategic planning as discussed in Chapter 3.

Requirements, often resulting in a specific deliverable or ability, are typically solicited from users. They, too, will guide the work, although at a more specific and deliverable level. Requirements can be considered the basis or "drivers" of tactics, that is, what exactly needs to happen and often how. Requirements will form the basis of organizational, process, and technology changes, and will dictate the scope, schedule, and budget of the program. Lack of clear requirements is a leading cause of project failure. Conversely, proper requirements are a leading factor in project success.

2.5 Strong Program and Project Management

Successful program and project management require skill and experience, not just within the technical realms of the effort but in the arts of resource management, scheduling, consensus building, problem resolution, contingency planning, and a myriad of other skills.

A common practice is to place the local technical expert in charge of a program or project because they are more familiar with the intricacies of the technology. Although this knowledge is needed, experience has shown that roughly one or two out of five technical experts are successful in the realm of project management on their first engagement. This is neither intended as a criticism nor to say that technical experts cannot perform as project managers, rather, that they must have the tools, training, and desire for this different role. The more critical the program or project, the more experienced and seasoned a selected project manager should be. A program and/or project manager is ultimately responsible for delivering a business improvement within the time and money allocated. The decisions and actions needed are often not technical in nature, but managerial and leadership based.

2.6 Scope Control

Scope control can be simply defined as working on those things needed to deliver program requirements. "Scope creep," or the gradual, often undetected and unfunded addition of functionality and desires, will likely affect a program near the end of the schedule when it is too late to recover without significant action being taken. Scope control has many facets, including

- Clearly stating requirements up front (what will be done and what will not be done);
- Keeping the scope as minimal as possible;

- Breaking efforts and projects up into smaller, more manageable deliverables;
- Keeping close tabs on work efforts and curtailing any expenditures that are not directly related to deliverables; and
- Identifying where new scope is actually needed and formalizing appropriate actions to secure related changes in schedule, budget, resources, and expectations.

Scope control is critical to program planning, both as an upfront tool for appropriately communicating plans and resource needs and an execution monitoring and control tool to help ensure program success.

2.7 Drivers and Constraints

Drivers are the motivations or pressures that cause one to do something; drivers are typically the cause of one's objectives. A utility will typically not implement a program plan unless there is both a strong motivation and a justification. Such motivations, or drivers, can come from either inside or outside of the organization. Internal drivers could include such problems as the inability to manage massive amounts of data, lack of user interaction, or the inability to collect key metrics and make associated business decisions. External drivers could be public pressure to track O&M costs, the need to meet regulatory requirements, or customer dissatisfaction with publicly advertised information. As a first step in planning, it is important to clearly understand the drivers involved in instigating formation of a program plan. These drivers often shape the objectives and requirements of the program.

Constraints are factors that either limit what is done or cause it to be done in another way. Constraints can be either internal or external; in either case, they typically make a job more complicated. Internal constraints include staff skills and availability, lack of technology infrastructure, or limited funding or time. External constraints may include political or regulatory deadlines and commitments or public relations and reporting challenges. Such constraints create additional work and hurdles that must be overcome; as such, they affect the scope, schedule, and budget of a program. Part of the job of a program manager is to identify and remove, minimize, or plan around as many of these constraints as possible, thus saving money and reducing project risk.

It is important to capture and understand both drivers and constraints as they will shape program plan activities and execution and are the ultimate basis for program expectations.

2.8 Understanding Current and Future States

A plan implies that something will be done or some change will be implemented. A plan is a description of the steps and resources involved in getting from one condition or state to another. To determine the steps needed to get from here to there, the definitions of the *current state* (i.e., as is) and the *future state* (i.e., to be) need to be completely and accurately understood. The program plan is then, effectively, the method and actions derived to get from a current state to a newer one.

Although understanding specific program details is important, there needs to be a reasonable limit to how much detail must be collected before action is taken. Indeed, "analysis paralysis" can lead to setbacks because of a perceived lack of progress. This is part of the art of developing a program: knowing when there is enough critical mass of information to take action.

To what level of understanding then should as-is and to-be states be understood? Whereas the initial level of understanding will need to be conceptual, the eventual target for understanding is at the business process and data level. Data are the key to most business decisions, and business process defines what is done with the data to get to those decisions; information technology is the suite of tools used to process the data for required purposes.

If the needed data can be defined, how it will be obtained and what will be done with it (i.e., to be) as compared to what information there is, how it is currently obtained, and what is currently done with it (i.e., as is) are key to determining work scopes. What needs to change can then be defined. This is often called a *gap analysis* and is a critical element of program planning. Once what needs to change and why has been defined, the program manager can then define how, where, when, and who will change it. These elements represent the basis of a fully prepared program plan.

2.9 Business Environment

Whether the plan is to install a new supervisory control and data acquisition (SCADA) system or a new accounting process, it is part of the business of the utility and, thus, part of the utility business model. Because the program plan will have something to do with the utility business, it is necessary to understand how it fits in. The following topics discuss several factors critical to development of a program plan that either fits within an existing environment or creates a new one.

This section covers important business and IT issues that, although may not be a part of the final program plan document, need to be understood by the program manager to properly plan and execute a program.

2.9.1 *Prime Interrogatories*

When approaching a program as a whole, some basic questions need to be asked. The "prime interrogatories," or first questions, are a great place to start (i.e., who, what, when, where, why, and how). Although these questions have been around a long time and may seem obvious, Zachman (2006) applied them to a methodology for determining IT architectures and, by extension, the plans to implement them. If these questions are not answered, a complete problem or solution statement will not be obtained; by default, this will result in an incomplete program plan and potential future difficulties. Prime interrogatories should, however, often be asked in a different order (i.e., what, why, when, how, who, and, finally, where). What are we doing and why? When is it due? How is it going to be done? Who has those skills and who will be responsible? Where will it be? Although some of the answers, such as who, may be a given, these questions should be asked of every problem, large or small, at increasing levels of detail.

2.9.2 *Business Modeling*

The full concepts of modeling are too in-depth to discuss in this chapter. However, some case examples are shown here to illustrate the following point regarding program planning: that modeling helps capture the essence of problems and solutions and helps to standardize communications for all parties involved. Ultimately, this is a form of communication and is needed to firm up the essence of the problem/opportunity of the as-is state and the definition of the to-be state, all of which are the basis of eventual time, budget, and resource needs.

Ideally, an IT system will mirror the business system it supports. The tools and processes used should follow those of the business, as in the adage "form follows function." To determine the steps needed to create a program, there must be an understanding of these business processes. This is best done via simple modeling; laying out schematics of the business process, who is involved, what data are used, how it moves, and so forth so that people on both sides of the business/technical fence can understand and agree on it.

For practical planning purposes, modeling can be just a picture or schematic of how something works. In basic terms, it can be broken down into fundamental building blocks of input, process, and output, with such steps linked together as necessary to show a general concept. Modeling can (and should) be done at high levels and, eventually, on the technical side and all the way down to detailed data levels. In Figure 4.1, a simple communication level model shows how "on-the-ground" operational data feed operational systems or processes. The model also shows how these systems share information. This is useful, for example, in showing the finance group how SCADA data eventually support financial planning.

A more conventional model of an overview "business context diagram" for a water utility is shown in Figure 4.2. This models the basic business and associated conceptual data flow.

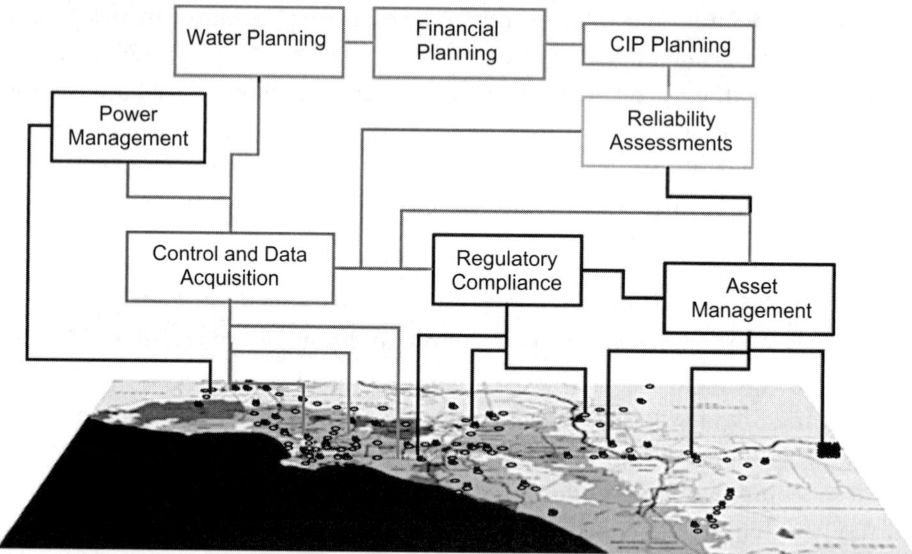

FIGURE 4.1 Sample of a high-level conceptual model of business functions and field data interaction (courtesy of D. Henry, P.E., MWD of So. Ca).

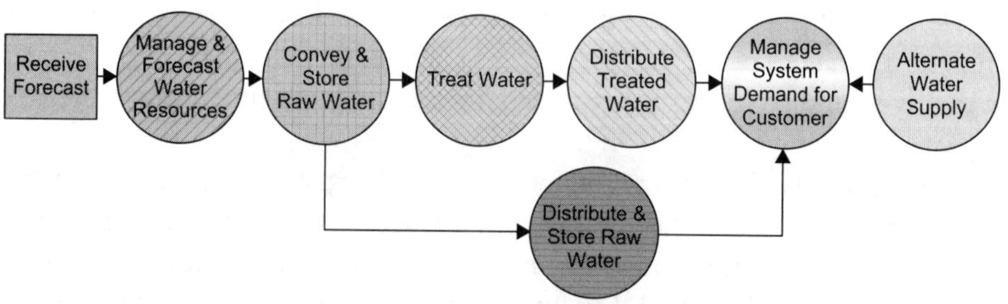

FIGURE 4.2 Sample business context model (courtesy of D. Henry, P.E., MWD of So. Ca, in concert with EMA Engineering).

In keeping with the concept of elaborating on models to increasing levels of detail, Figure 4.3 shows an expansion of each of the basic functions shown in the preceding figures, but at a more technical level that begins to allow for program planning.

This modeling can, and eventually should, be expanded and elaborated on until a complete as-is model is created to the business process level. At that point, fixes to the as-is state and refinements and additions to the to-be state can be identified with some intelligence. In yet another example that includes personnel, applications, and processes (the triad of people, process, and technology), the "swim-lane" model shown in Figure 4.4 models how a particular system will be designed.

As these models continue to evolve, typically during program execution, it will be found that there are more sophisticated, automated, and interactive modeling tools that are available to, and used by, software programming staff.

If these more formal modeling processes are to be undertaken, it is recommended that a pilot study be performed to better understand the effort and nuances involved. A process should be selected that is in need of an initial IT system or upgrade, but one that is not too large or too critical. If necessary, a professional should be contracted who can both facilitate the harvesting of business process and data knowledge from staff and be able to translate that knowledge into the levels of information needed by technologists for development. The process itself is typically broken down into steps, each with associated levels of detail and deliverables. For a singular application, these steps are as follows:

- Business objectives,
- Conceptual business process diagram,
- Logical process diagram,
- Data map,
- Function diagrams,
- Programming specifications, and
- Physical specifications.

There are many ways to model something and numerous formal schools of thought on how it should be done. The more important point to realize is that these models, in whatever form, create a tool for communicating and coming to consensus among all the different levels of a program team. They also illuminate the following:

- Duplication of data,
- Missing data,

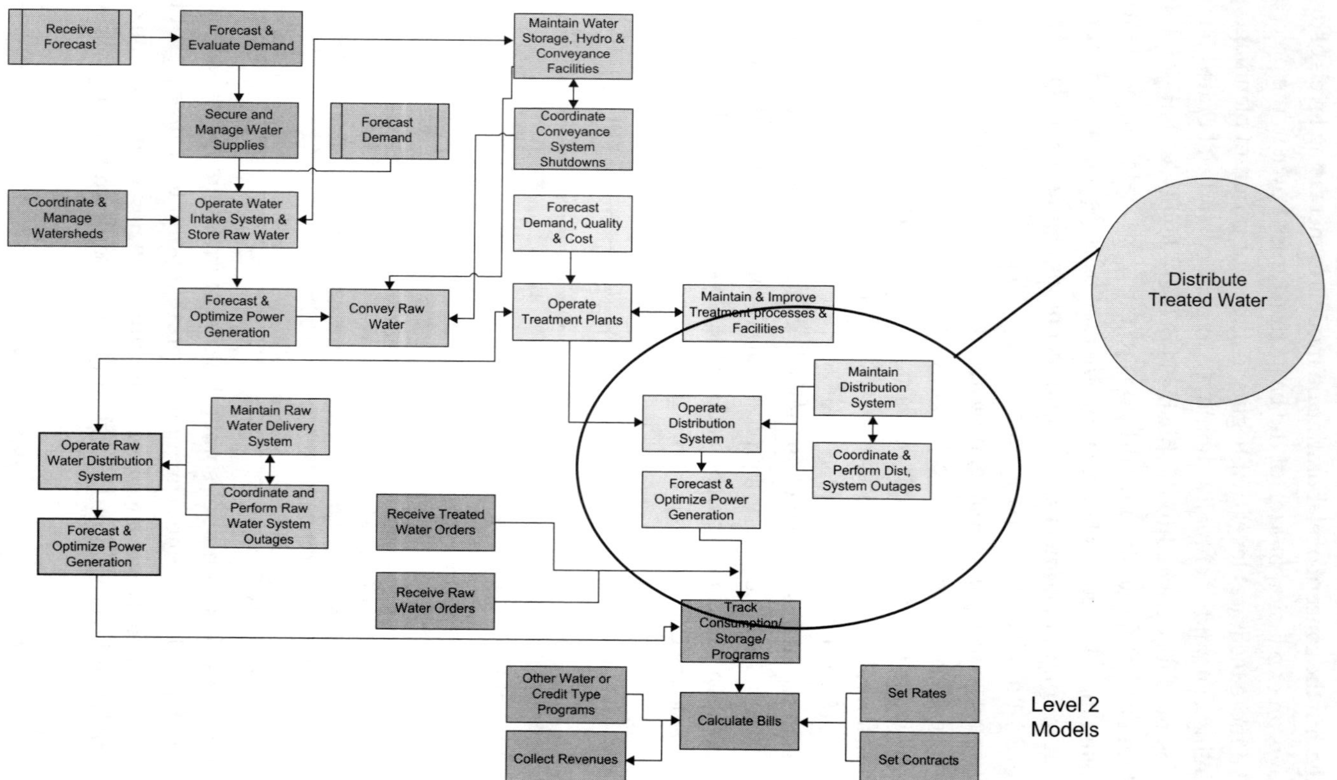

FIGURE 4.3 Sample core business function model, expanded from Figure 4.2 (courtesy of D. Henry, P.E., MWD of So. CA, in concert with EMA Engineering).

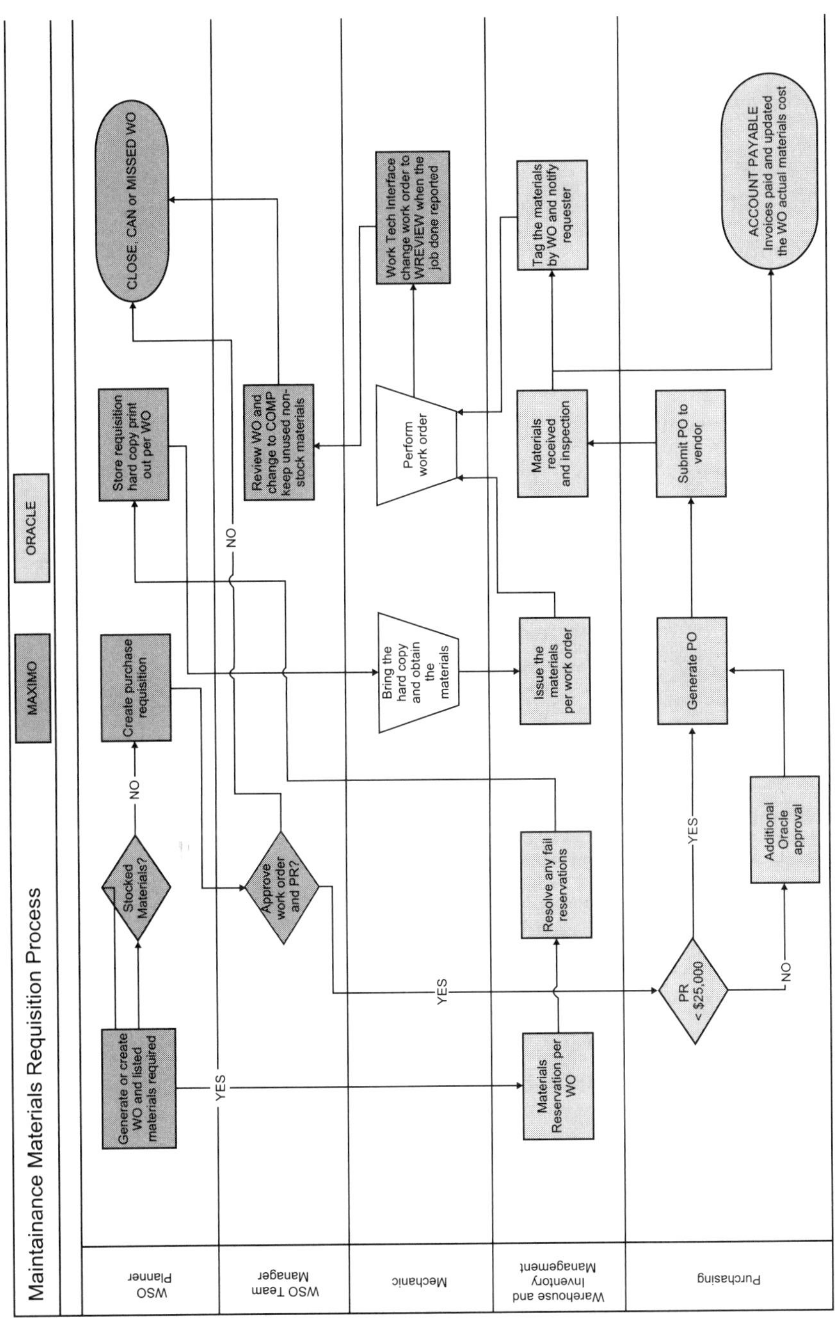

FIGURE 4.4 Sample swim-lane model for a requisition process (courtesy of S. Ma and D. Henry, P.E., MWD of So. CA).

- Data being generated for no apparent reason,
- Duplication of applications,
- Missing applications,
- Missing interfaces between applications,
- Duplication of business processes, and
- Missing business processes.

Most importantly, models allow the program team to understand what they are dealing with now and what needs to be done in the future. Only when there is such an understanding, with or without models, can a program be properly planned, staffed, funded, and eventually executed.

2.9.3 *Managing Complexity*

If an enterprise is ever fully modeled, the magnitude of the complexity the utility really has might be astounding. It may also be more complex than it needs to be, typically as a result of years of add-on, stand-alone, and ad hoc integrated systems. However, to understand, minimize, and manage the complexity of a program, some level of modeling, as discussed previously, is likely necessary.

A key cautionary point to bring up, and one that should be conveyed to the plan's audience as well, is that *the plan is not creating this complexity, rather, it already exists; the plan is intended to manage and even simplify it.* Do not let the presentation of existing complexity be confused with what is being proposed. The audience may simply think, "This is too complicated and there is no way we are doing this." Although a less chaotic future is needed, it should not overwhelm the audience. The objective is to streamline a system and make it a viable path for the future. As Figure 4.5 demonstrates, this complexity, even at a conceptual level, can be daunting (the figure is based on actual utility data; further elaboration on the development of this is provided in the case studies in Chapter 10). Figure 4.5 represents an IT architectural "framework" in the most literal sense. Further development of this model, which is currently underway, will link data to specific assets that generate data, thus creating a merger of the mutually dependent physical/operational/geospatial and the conceptual/business architectures.

There are pros and cons to modeling and managing the complexity of a utility business and the mirrored IT architecture. Despite the difficulties, however, it is a task that should be undertaken to help ensure program success.

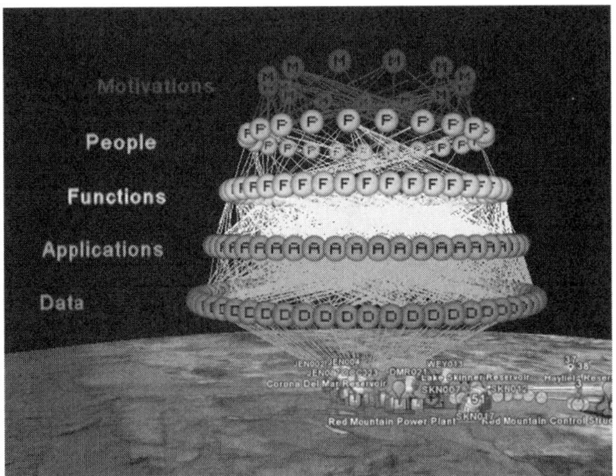

FIGURE 4.5 Utility business architecture framework: a picture of complexity, even at the conceptual level (courtesy of D. Henry, P.E., MWD of So. CA).

On the negative side, complexity management

- Can be difficult and time-consuming to fully capture;
- May embarrassingly point out gaps, redundancies, and problems in both process and release of the product;
- Can be overwhelming, and may turn off upper management's reception of the construct; and
- May result in identifying additional obvious changes that are needed, thus instigating span of control and responsibility issues. Such changes could be in organization, business process, or technology:
  - Organizational and process changes are often not under the span of control of IT, and thus create a potential challenge (executive management may have to be carefully involved to resolve issues); and
  - Information technology/technology changes, by themselves, may not be effective unless organizational and process changes lead the way.

On the positive side, complexity management

- Provides a reality check for possibly several years of business and IT sprawl (akin to urban sprawl);

- Shows inherent connections between business and technology, a critical bridge that is often overlooked by both business and technology camps;
- Identifies, specifically and quantitatively, gaps and redundancies in business process, data, organization, applications, and architecture;
- Provides an as-is picture of the utility from which to effectively determine a to-be state, which is the basis of the program plan;
- Allows for the ability to prioritize actions and to segment scope based on defensible business drivers;
- When simplified, it can provide a universal communication tool for all parties involved;
- If done correctly, it can, and should be, the objective basis by which to launch and defend planning and execution;
- Helps define where systems are interconnected and where associated data can be shared, which helps minimize redundancy in process, data, and data maintenance and allows for "one version of the truth" in business data; and
- Can be simplified or segmented to show only a specific or localized part of the objective, organization, application or data flow, while still maintaining relevance to the whole.

In planning such modeling, it is wise to consider whether to model only a portion of the system or the entire business. Modeling only a portion of the system is a much more manageable task in the short term, but may lack several internal and external influences and the ability to foresee longer-term issues. Modeling the entire system while generating a big picture can be a significant undertaking and may delay fruitful results if release is deferred to the end of the process. A suggestion is to prioritize business processes and model them in sequence, releasing them as developed, with the caveat that they may change as the effect on them by other systems is discovered.

Part of the objective, if there is a holistic approach to the program, is not to add yet another set of processes, applications, and databases and end up creating more unchecked complexity; rather, it is to manage and even simplify the complexity a program already has. A potential problem, however, is that, in many cases, existing systems were never engineered to work together as part of an overall business support model. Whereas the scope of efforts may not be to perform such a global task, consideration of future architecture can help prevent the problem of simply adding another system and more complexity. It is important to try to understand where business objectives, processes, applications and, specifically, data intersect and are shared. Then, it is important to try to use the appropriate technologies and tool sets so that data and processes can be shared as progress is made.

2.9.4 *Parties Involved: "The Big Three"*

In all aspects of program planning, three fundamental groups of people will ensure either the success or failure of the program plan. These are management, customers (including employees, who are often considered the customers of software projects), and service providers. In general, both the program plan and program manager must engage, align, and maintain all three groups during both the development and execution of the plan. It is important to realize that although these groups are necessary for the success of the program, they can also represent sources of difficulty, each for their own unique reasons. For example, management may either sponsor a plan or cancel it for strategic or political reasons; customers may demand a product or service, only to reject it if it is not released as expected. Finally, service providers may advocate a change and the associated work, only to reject it or fail at the method of the change.

2.9.4.1 Management and Constituents

Management typically has ultimate accountability (both financial and overall) for a program, and, as a result, is motivated by broader business decisions and objectives. Management wants to be able to declare positive developments to their constituents, whether this is a board of directors, the public who they are entrusted by, or some larger governing body. One can generally address management's concerns with the following frequently asked questions (FAQs):

- What is it?
- Why are we doing it?
- What if we do not do it?
- What are others doing?
- How much will it cost?
- How long will it take?
- What is the risk?

- What will be different when it is done?
- Who is involved?
- What are the options?

If these questions cannot be answered, the plan may not be mature enough to propose to management and further refinement may be necessary.

As a side benefit, these same questions and answers form the basis of a useful handout or brochure on the program, one that can be given to board members, managers and staff, the public, or other interested parties. This document, concisely and neatly packaged as a primer of sorts, should be easy enough to digest that it can then be discussed or repeated by managers without further reference. The benefit of this is that by being able to talk about the program, in business terms, management will begin to take ownership in the program, thus making the effort part of the culture and helping to ensure progress. Such a package also helps ensure consistency of the message throughout the organization.

2.9.4.2 *Customers*

Customers can be either internal staff, such as the accounting or engineering department; external, such as the customers of the utility; or sometimes both, as with a new billing system.

External customers, typically the public or another agency, also want to know what is being done and why; after all, they are generally paying for the effort and want to get their money's worth. As with managers, they want to know the answers to the same FAQs as those previously presented. Customers also must express either a need for, or concurrence with, a program plan to proceed.

Customers must also be willing and able to support and adjust to the changes that will happen. It could be that the changes are ultimately transparent or it may be that significant change from what was previously the norm will cause some upheaval and challenge. Change management, which will be discussed in section 2.10.7 of this chapter, becomes important here, as does piloting, education, and training.

2.9.4.3 *Service Providers*

Service providers, or those entities that will deliver the actual technology, can be internal or external and are likely to present a significant planning, expense, and complexity factor. Depending on the nature and size of the organization, either internal skills are available or the work will have to be contracted out.

Hybrids of this service strategy often occur as well, with portions of the work done internally and portions done externally; however, this should only be done with caution along with an excellent understanding of the details and relationships involved. Either internal resources, external resources, or a hybrid combination of the two will be needed. Ultimately, this is resolved through resource assessments and careful contracting.

2.9.5 *Services Environment*

An often overlooked cost is the environment that may have to be created to allow a program plan to be executed. The program manager will need to assess the environment needed to provide IT services, both during program execution and after release of the program, and provide what will be necessary. As such, space, software, hardware, and tools may be needed for the following:

- Project managers,
- Project administrators,
- Contract administrators,
- Programmers,
- Database administrators,
- Communications,
- Security,
- Network,
- Hardware,
- Field technicians,
- Quality assurance/quality control,
- Testing, and
- A help desk /post-release assistance and debugging.

Although some of these resources will not be necessary after the peak load of program or project development, many of them will also be required for maintenance and help-desk purposes. If these resources are not available in-house, methods to obtain them should be planned and executed. It is important to be aware that external contract staff and consultants can cost more than internal staff; cost is also a factor that fluctuates widely with the economy and supply.

It is important to include not just the software that will be used, but the software that will be needed to develop and maintain the new system. Specific discipline experts should be able to identify these needs, which may include the following:

- Modeling software such as Rational Rose (IBM Corporation; Costa Mesa, California), Enterprise Architect (Sparx Systems Pty. Ltd.; Creswick, Victoria, Australia), and similar systems;
- Coding software such as a .NET (Microsoft, Redmond, Washington), Oracle (Oracle Corporation, Redwood Shores, California), SQL (Microsoft, Redmond, Washington), or similar products;
- Report and interface building tools;
- Security administration tools;
- Testing tools;
- Version control and tracking tools;
- Communications configuration and monitoring tools;
- Database administration tools; and
- A test or prototyping setup and so forth.

The typical progression of development to release phases includes the following levels of hardware (some of which can be economized):

- A development environment or sand box, where programming is done;
- A stage environment, where a complete turnkey replica of the final system is created, tested, and debugged. This can also be used to test future changes and post-release upgrades before going live;
- A production environment, where the final product is released to users. Be sure to consider redundancy and backup systems for critical applications and data; and
- "Lab" setups where, for example, instrumentation to communications to data collection and processing can be tested. This can also serve as a prototyping or pilot-testing environment.

It is important to consider these resource needs during planning and budgeting. In addition, lease options should be considered in situations where you expect specific hardware will not be needed after release.

2.9.6 *Utility Business Processes and System Options: How They Affect the Program Plan*

This section discusses the often unique nature of utility business and, as such, the unique nature of program solutions. Because of the singular nature of a utility, scope, schedule, and cost are often not the summed cost of installed off-the-shelf packages, but rather of custom-built or combined environments.

Two extremes are present when implementing utility business systems. The system is either custom designed to meet the existing business process or an application is installed and the business process must change to work within the constraints of the application. There are also hybrids in between to fill the spectrum. In IT terms, these application extremes are called *custom applications* or *commercial-off-the-shelf* (COTS) systems. Hybrids are termed as custom applications with some commercial components or COTS with some customization, the latter being the most common for generic business applications, such as commercial accounting systems customized and configured to meet internal utility requirements. Additional possibilities and variations on this theme are software as a service and other Web-based and remotely hosted applications, which are often developed and managed externally by third parties on a fee-for-service basis.

Whereas the water and wastewater industry does have the standard "any business" process of accounts payable and receivable, human resources, and so forth that can be met by COTS products, utilities also have some unique core business processes that are not met by COTS. *Core business* is defined as business functions that are performed to meet mission statements. For example, utilities for the water industry plan, procure, convey, store, treat, meet regulations for, and distribute water. The utility may also engineer, maintain, and control systems and even generate power. Utilities do this because they are a water or wastewater agency, not a car dealer; the actions they take are the result of the utility's core business. Each of these core business functions typically has a custom-built IT system in place to support them. It is common for these processes to be unique; therefore, the applications, processes, and systems to support them are unique as well (i.e., either custom-built or highly customized COTS systems).

Given this variability and uniqueness among water and wastewater utilities, it is critical to understand the specific nature of each business process if an effective system is to be created to support it. The benefit of a cookie-cutter IT architecture that a chain of grocery stores may leverage is not available to the water or wastewater industry; water and wastewater utilities do not yet benefit from these economies of scale. Regardless, it is incumbent upon utility IT management to understand the business process itself to both create and maintain any effective system, whether it is a COTS, custom application, hybrid, or a combination of these. The consequences of not having this understanding are to spend both capital and O&M money and political capital on a system that may be challenged.

2.9.7 *How Data Affect the Program Plan*

Although data are covered in more detail in Chapter 2, some understanding of the technical challenges caused by different types of data is necessary for proper program planning.

Data are at the root of most business processes and decisions. The typical evolution of information is as follows: from data to information, to knowledge to wisdom, and then to decisions and action. Without data, most decisions cannot be made in a wise manner. What happens between data and resulting decisions is dependent on process. Without proper knowledge of the data and the process, systems cannot be properly created.

One of the greater challenges of program planning is understanding that, generally, no single IT system provides all the data needed; data have to be integrated from several systems to create useful information. For example, how do you minimize the cost of treatment? One must know current treatment variables that may come from weather, SCADA, and laboratory information systems and the current cost data, which may come from purchasing, accounts payable, energy bills, labor, and maintenance records. Data must then be assimilated from these disparate systems and intelligent and integrated information created via some logical process. A software system is often used to automate the process, either for a one-time analysis or for an ongoing metrics-type monitoring "dashboard" or regular reporting process. The point is that data from different systems often have to be integrated to become useful. Planning for the associated systems and interfaces is crucial to proper program resource, schedule, and budget allocations.

2.9.7.1 Different Types Require Different Treatment

A correlated challenge to the disparate source of data is the disparate format of this data. Data are stored, manipulated, processed, and extracted in many different ways; in addition, much of these data are not easily made compatible (this is especially true of real-time data being compatible with transactional data, as defined in the following sections). Having to build systems to accommodate blending data can affect costs; this needs to be recognized during the planning phase.

2.9.7.2 *Transactional Data*

Transactional (e.g., financial or customer) data are the type of data typically found in accounts payable or payroll. They represent a trans-action-based set of data used to record data about a specific, singular transaction, typically found in financial or business database applications. These data can be stored as relational, star schemas, cubes, and so forth.

2.9.7.3 *Time-Series Data*

Time-series (e.g., control or monitoring) data are typically related to physical field processes and are conventionally associated with control systems. These data are based on conditions at a snapshot in time (vs a transaction), typically repeated over time at regular intervals. These data can be stored in relational systems, although this is not always efficient; it is easy to store, but difficult to recall. Because of the format and quantities of time-series data, proprietary databases that use filtering and compression are typically used to manage these often massive amounts of time-series data.

Time-series and transactional data are often obtained, formatted, stored, manipulated, and retransmitted in different ways, making the merger of the two types of data difficult and, often, less than optimal. In addition, from both business and technical perspectives, these two types of data lie in two different disciplines: business transactions and process control. However, operational control data are the product of the central nervous system of the utility; they report flows, power consumption, chemical use, production levels, and so forth, which represent the metrics of the core business, and should be recognized for their value.

With such a wide gap between transactional and time-series technologies, the merging of these two different sources of data and information has been a universal water and wastewater utility challenge. These technology challenges have not been fully solved by industry, and the ability to understand the core business links between transactional and real-time data—to build an effective bridge—have limited the ability to achieve one of the greatest potential efficiencies at utilities, that is, linking operational metrics to business decisions.

In summary, it is important to carefully examine the data requirements, sources, structures, and integration needs and opportunities because the program scope and budget may be heavily dependent on these factors.

2.10 "Soft" Operations

Gaining agreement on a program plan manifests itself in a spectrum of activities, from qualitative and intangible interpersonal relationships to factual, logistical, and

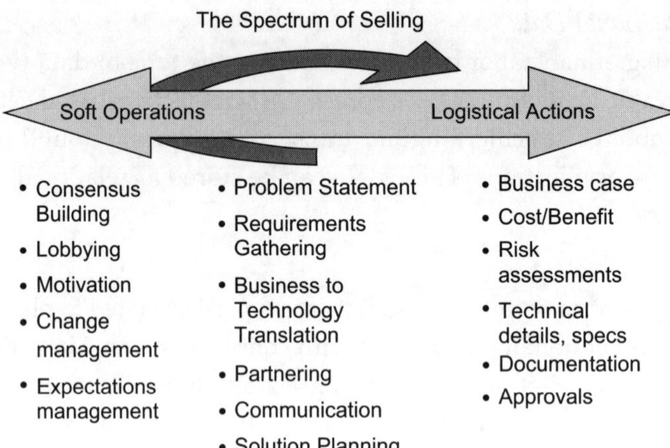

FIGURE 4.6 Selling the program plan (courtesy of D. Henry, P.E., MWD of So. CA).

qualitative deliverables. One way of breaking down this spectrum of activities into manageable tasks is through use of "soft" operations and logistical actions, as illustrated in Figure 4.6. Planning activities generally start in the center column of the figure, then move to soft topics such as consensus building, then refine the center column, and, finally, deliver "hard" products on the right side of the diagram.

During development of the program plan, some interpersonal activities will need to take place. Soft operations are those that are interpersonal, have no formal deliverable or product, and are generally categorized as successful by whether or not agreement has been reached. Such activities include consensus building, lobbying, expectation management, leveraging proper motivations, and change management.

The efforts and pitfalls discussed in the following sections provide some guidelines of what needs to be done as part of preparing a plan that will meet everyone's needs. The program plan is not just about a document; it is about getting everyone moving in the same direction toward a new state of business and arriving there successfully. The document merely provides the agreed road map, deliverables, and timeline that can be referred to during execution.

2.10.1 Consensus Building

Building consensus, at both a strategic and tactical level, is critical to a program plan. Although a plan may be mandated down to staff, it may not be supported by them. Similarly, although a system may be handed to customers, they may reject it. Both instances can result in failure. In the context of program planning, *consensus* can be defined as when a critical mass of support is obtained to the level that an effort becomes a "pull" vs a "push" (i.e., where customers are pulling for the efforts vs the efforts being pushed on them).

Referring back to "the big three" (i.e., management, customers, and service providers), the amount of work needed and the resistance to overcome to implement a plan is significantly higher in a push vs a pull situation. The challenge is to educate and facilitate understanding between all three parties on the problems, solutions, and workable path that all can agree on and move toward.

However, airing existing system problems or pointing out a deficiency in someone else's sphere of control can inhibit progress. Therefore, it may be prudent to let customers take the lead; in the best situations, the customer will make both the problem and solution statement (ideally, with some background facilitation). This approach may actually be good for the program as it becomes "business process X" project sponsored by the business owner instead of "IT project Y," and may get more support that way.

Instead of approaching consensus building from a "problem" perspective, which is essentially advertising a negative situation, reach for positive opportunities. Although a solution may be the same in either case, how it is packaged may make a huge difference in the level of support. For example, instead of asking, "How do we solve this problem?" ask, "How do we achieve this benefit?"

The challenge of consensus about the problem also exists for the solution. It will take some back and forth, but agreement needs to be complete, clear, and concise in the end as the agreed upon solution, or to-be state, will be the basis of the scope, schedule, and budget. Once there is consensus between involved parties, the plan is well on its way to formation and execution.

2.10.2 *Lobbying*

Depending on the level of influence and authority immediately available to a program manager, it may be necessary to secure additional support and cooperation. Indeed, rarely will one individual define organizational problems, solutions, and plans and also get the whole program funded and rolling. To gain the additional support needed during consensus-building efforts, it is important to identify decision makers, those who will be affected, and who in these areas have the needed influence and authority. Discuss what this plan means to them and the organization and try to secure their support. If their support is not obtained, understand why and either negotiate a solution or redefine the plans and objectives until support is obtained.

The objective here is to build a group of supporters that will, in turn, lobby for the program based on its merits.

2.10.3 *Partnering*

Partnering is a concept that comes from the construction industry. The experience in that industry was that the owner and the contractor never met until they went to court at the end of a project to settle a dispute. Realizing that the only winners in this situation were law firms, owners and contractors began to meet at the beginning of projects to discuss what they both did and did not want out of the process. In effect, they became partners in mutual success.

It is extremely helpful to establish a high level of give-and-take between parties with a common interest rather than a fight over deficient contract language between self-interested parties. It is strongly suggested that partnering be considered in program and project efforts, specifically for those with significant contracts. Partnering is also highly recommended in completely internal settings. Partnering consulting, facilitating, and training services are available and are recommended for first-time efforts.

2.10.4 *Managing Communications*

It only takes one misunderstanding to set a plan back. Therefore, it is important to make sure to communicate thoroughly, consistently, and regularly to all interested levels of the program. In addition, a master presentation on the program or project should be prepared that can be selectively updated and customized for each audience (e.g., by not showing certain slides, such as database details, to executive management). Similarly, a consistently formatted monthly report for status updates should be prepared.

One streamlining approach that seems to be appreciated by the audience and project managers alike is preparing and presenting the monthly report in presentation format vs a text document. In addition, as discussed previously, a FAQs file for handouts should be prepared. By communicating regularly and consistently, confusion and misunderstandings will be limited.

2.10.5 *Managing Expectations*

Each person even remotely involved with the program plan and deliverables will have a different perception of what is being done, what problem will be solved, or what opportunity will be delivered to them. Problems may arise when this variety of expectations is not met. This realization may happen midway through development or even after the product has been delivered (i.e., when the money to fix it is gone). It is important then to set expectations clearly and appropriately and reinforce or manage them throughout the life of the project and program. Keep expectations within a reasonable realm and be cautious about people over-interpreting what will be delivered.

In terms of managing communications, it is important to be clear about what the program/project will do and present that intention in simple and universal terms. It is equally as important to be clear about what the program/project will not do. Finally, it is important to keep scopes minimized and to keep published expectations slightly lower than what is planned to be delivered; this may help actual results to be higher than expectations.

2.10.6 *Motivation*

When working with individuals during the processes of consensus building, lobbying, planning, or execution, it is important to understand that each person will have differing motivations to support or not support the plan. This can be as much a personality trait as it is a function of role and responsibility. William Marston (1928) identified orders of motivations in people, and these are adapted in Figure 4.7. These

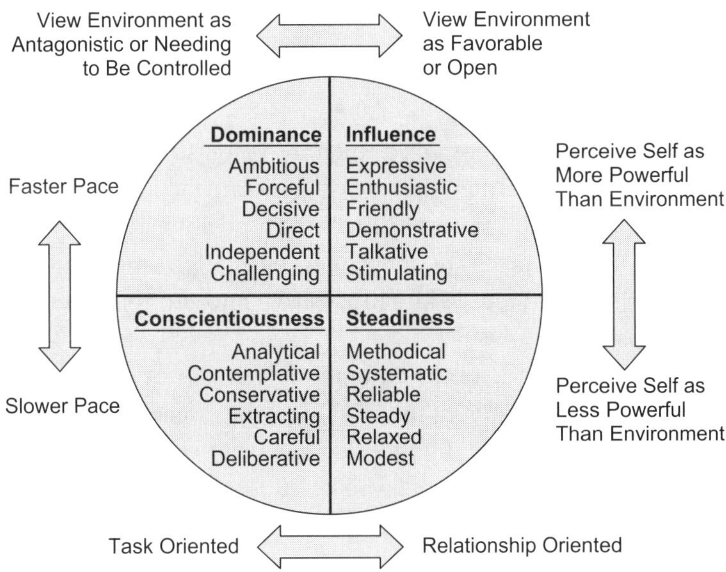

Figure 4.7 Understanding motivations: DiSC® (Marston, 1928).

may be helpful to be able to identify what motivates certain personalities in an effort to persuade them to support the plan. As a program manager, present your arguments accordingly. It may also help in organizing the program team by placing individuals in positions where their motivations are best rewarded.

2.10.7 *Managing Change*

Change, for the sake of program execution, generally occurs at two levels; organizational and individual (although enough individuals can affect the organizational reaction). If changes are either too great or too quick to be adapted to, rejection will occur. Understanding how change can affect and be accommodated by both categories will be important.

For both categories, Gleicher and Bechard (1969) developed a simple formula, slightly modified here, as follows:

Change happens when

$$D \times V \times P > R \tag{4.1}$$

Where

D = dissatisfaction with the current state,
V = a vision of a future state,
P = positive first steps, and
R = resistance to the change itself.

To affect this change, one may need to either increase the left side of the equation or reduce the right side. Although increasing dissatisfaction is not recommended, setting a better vision for a future state and taking a positive first step can be an easy start to the process. A pilot program or a prototype is often small enough to be cost-effective and low-risk first steps while illustrative enough to successfully prove a concept.

Reduction on the right side of the equation is a matter of dealing with concerns about the upcoming work involved, the risks, and the effectiveness of the proposed new state. This resistance is generally reflected as fear or concern, and these concerns exist in different conditions at different levels of the organization. They may be cost or credibility based at the executive level, assignment based at the service-provider level, or roles and responsibilities based at the customer level. It is important to note that this resistance is generally not about technology; rather, it is about culture and vision.

At the organizational level, change is typically a question of logistics and how much an organization can accommodate change while still performing its day-to-day functions. It is typically a question of how many resources the organization can spare to affect the change. If there are not enough resources, either more resources need to be brought in or the change needs to be throttled back to an acceptable pace. This condition is not necessarily limited to technical resources; it may be that the availability of legal contracts, purchasing, operational, and customer resources are as likely to cause delays as the lack of programmers. This effect is typically a function of project life cycle, timing, and magnitude, and depends on what stage the program is in and where the resource bottlenecks are created. Therefore, a program manager should plan and communicate accordingly and manage scope and expectations, as discussed previously.

3.0 DEVELOPING THE PROGRAM PLAN

Having covered the basic variables of program planning, and in line with the program plan outline presented earlier in this chapter, it is now appropriate to discuss developing the actual plan. At this point in the process, executive support, user involvement, stakeholder alignment, lobbying, and change management should be taking form and initiated. There should be a consensus on the objectives, drivers and constraints, problems and opportunities, specific requirements, and, finally, expectations. The business, process, and technical environment as-is and to-be states should be understood. A gap analysis and general options on how to achieve the to-be state should be outlined and agreed upon. Communications should start taking form to help achieve a consistent message to involved parties.

At this point, there should also effectively be agreement on the plan. What remains is merely a formalization of this understanding into a document, including some details on how the plan will be executed. The following discussions elaborate on the specifics of major program plan documentation deliverables.

3.1 Defining the As-Is State

The as-is state is likely in some form of need or it would not be a target for upgrade, replacement, or automation. Alternatively, a new state may present such strong opportunities that even a functioning current one can be left behind. In either case, the current state must be clearly understood if changes to it are going to be planned. The current state should be modeled, at only a high level if necessary, and the faults or opportunities identified. The effects of these faults (i.e., cost, risk, noncompliance, hours, rejections, and so forth) should be quantified; in other words, what are the faults of the current system costing? Provide any helpful qualifiers as well. Whereas opportunities may reside in the to-be state, they imply some basis of change to the as-is state and, therefore, require some discussion here. Where possible, opportunities should be quantified and qualified as well. Both problems and opportunities should be expressed in terms of people (i.e., organization), practice, technology, or a mixture of these, as appropriate.

In developing the program plan, language should be kept in business and layman's terms and technical support should be provided as appendices. What needs to change and why should be clearly stated, making sure there is consensus from process owners on the definition of the current state, its faults, and opportunities.

3.2 Defining the To-Be State

There is some interplay and iteration between the to-be state and requirements. In one sense, the requirements cannot be formalized until a to-be state is defined. In another, the requirements define the to-be state. The general sequence is that this iteration begins with a conceptual understanding of what the vision for the future should be (i.e., what people want) in fundamental terms. From this conceptual vision, specific requirements can begin to be formalized. As these details are worked out, there will likely be some change and refinement of a more specific definition of a to-be state. It should be noted that as people learn more about what they want and what can be done, this iterative process of change and refinement will likely take place throughout the life of the program. Some flexibility and control to allow for this as a natural process should be planned and managed accordingly.

The program manager should work with process owners to formalize the new to-be state. Improvements and changes should be conceptualized and engineered at this time. Referring to the program plan outline, critical success factors (i.e., those factors that must be achieved for the program to be successful) should be determined (refer to Chapter 9 for a discussion of critical IT success factors). Risk mitigation and exception management should also be considered in this process. In addition, the program manager should identify how prior faults and opportunities are addressed. Along with stating what will be attempted, the program plan should clearly state what will not be addressed or delivered.

3.3 Gathering Requirements

Requirements are a necessity in any program or project plan. They provide the basis of all work to be done, deliverables to be achieved, and the associated time, budget, and resources required. Requirements also provide guidance and control for the program by aiding in identifying stray work expenditures. Requirements that are not captured, or clearly defined, may later result in either dissatisfaction by users for lack of functionality or they may affect to the program's schedule or budget by being included at a later date.

Program managers should write requirements down, in detail, as a formal document signed by the customer, service providers, and executive management. This document is the "contract" that states what will and will not be delivered. It need not be technical; conceptual may suffice. The program manager should stick to these requirements and refer to them often. If effort is being expended on something that is not in the requirements, it may be out of scope and consuming limited project time and money. If this "extra" task really does need to be in the requirements, then there is a change of scope and the program is entitled to a formal change order and associated adjustment to the budget and schedule, also signed and approved (if not requested, sponsored, and endorsed) by the customer.

Requirements need to be testable wherever possible, to a point that tests can be run to confirm delivery of the requirement. Therefore, in the end, the requirements should also be the basis of test plans. Because requirements are one of the first things generated and the test plan is one of the last things executed, requirements are the constant theme throughout the project or program. In addition, because requirements need to be testable to determine if they are met, they should also be quantitative (vs qualitative). It is easier to confirm specific functionality with provable results than to meet subjective and qualitative desires.

3.4 Translating Business Requirements to Implementation Plans

One of the toughest jobs in program and project planning is translating business requirements into personnel, process, and technical requirements, which eventually evolve into scope, schedule, and budget. This takes a particular skill set, that is, someone who can extract and develop the business model and then translate it into technical terms for the IT team.

The criticality in this lies in the fact that each business requirement may or may not have significant system effects. For example, the ability for one user at one location to see real-time information as soon as it changes at another location requires an entirely different and more complicated architecture than a batch-process architecture. In another example, users may need a specific data set, which happens to be located in a separate real-time database. This may require specific protocols, a unique data-mart, time-parsing and averaging, and so forth (both are real examples, the first resulting in the cancellation of the project because of the cost of implementing the required technology). These seemingly simple and singular requirements then have a significant effect on scope, schedule, resources, and cost.

It is important to realize that once the technical (or process or organizational) implications of the new system are determined, the costs to meet these requirements may be excessive. If this is the case, there may need to be some adjustment to, and iteration of, the to-be state. As such, this may be a good time to readdress requirements, scope control, and prioritization, as discussed earlier in this chapter. Finally, it is important to be aware that although business users may have little interest in understanding the details of these technology issues, they will need to be explained in business terms to convey the final scope of work and the cost and complexity consequences of these requirements.

3.5 Defining the Bridge from the As-Is State to the To-Be State

The bridge from as-is to to-be states is the basis of the actions to be undertaken in the program plan. These actions result in specific time, resource, and budget requirements. There are typically a myriad of options to bridge the gap from as-is to to-be states. Because of this variability, there is not room here to discuss every possible path. However, there are factors to consider that include cost–benefit analysis, risk assessments, business cases, and so forth that could be used to weigh each of the options. There are also likely preferences among management, users, and service providers that should also be considered.

Within the program plan itself, options that were evaluated should be discussed (in layman's terms, where possible). Discussion of these options should address organizational, practice, and technology effects and changes, as appropriate. Each option should have a pros and cons list, a capital cost and schedule effect, and, possibly, a total life-cycle cost analysis. Briefly describe how each option was evaluated and tabulate selection results. Show how the final selection fits together as a system to provide the to-be state and meets objectives and requirements.

Based on the selection of a specific option, the program plan should elaborate on the steps needed to get from the as-is to the to-be state. These steps could involve organizational or roles and responsibility changes, process reengineering, filling of various gaps, elimination of various redundancies, training, communications, and so forth. In many cases, the deliverables needed to execute the chosen option will result in additional specific requirements, which will need to be updated and communicated accordingly.

Steps in the program plan should be defined in terms of a schedule of tasks, preferably resource- and cost-loaded, and they should be correlated to have realistic estimates. These steps will be the basis of labor, resource, contracting, procurement, and completion functions, among others, and will form the basis of program plan activities. This can also be done for a single project or an orchestrated collection of projects. The program manager should consider using professional project management, planning, and scheduling tools here, especially for large efforts.

3.5.1 *Risk Assessment and Management*

There are two categories of risk to consider: (1) the risk that the project is intended to mitigate (a possible reason for the project) and 2) risks that occur in the project itself, which are called *project risks*.

Most programs and projects have some element of risk mitigation involved. The cost of risk mitigation can be measured in terms of insurance; that is, is the cost of doing this effort more or less than the cost of the risk of not doing it?

The reader should recall that risk is, conventionally, the *effect* of a failure multiplied by the *probability* that that failure will occur. This can be quantified if done carefully, and is regularly conducted as the basis of actuarial setting. Another factor that can be multiplied in is the ability to *detect* the failure when it occurs; if a failure is detectable, the risk may be lower than if compounding failures linger undetected. For example, if there is a known risk with a billing system, what is the probability that all monthly bills to be sent to customers will be wrong, what is the potential value of that error, and will it be detected before the bills are actually sent? Conversely, what is the cost to fix this?

From a project perspective then, how you identify, quantify, and value a risk becomes important in its comparison to the cost of mitigating the risk itself. This is another area where an independent party, or at least the owner of the risk, should buy into risk analysis and the cost to fix it.

Project risk is the probability and consequence that the project itself will fail to deliver on expectations. These risks are related to requirements, staff availability, scope changes, and many other factors. To deal with these, a program manager should do at least the following four things: (1) plan carefully, (2) prepare exit and contingency strategies, (3) set aside contingency funds, and (4) minimize and control scope.

3.5.2 *Business Case Justification*

A business case justification is a document that analyzes the pros and cons of a project and provides a case for moving forward with the effort. The need, opportunity, and justification for plan execution should be stated in this document. In addition, the document may be internal, external, or both. In larger organizations, a business case may need to accompany each individual project proposal. One way to categorize business cases is as follows:

- Regulatory mandate,
- Public/board mandate,
- Risk mitigation or recovery, and
- Business process improvement.

Most of the time, a program, or project, will be a combination of two or more of these cases, and can be presented accordingly (e.g., a risk mitigation project, with some business process improvement). Whereas a cost and benefit or return on investment (ROI) can be calculated for any of these cases, they may be negative for the first three cases; in addition, an ROI may not even be required because of their underlying drivers. A business process improvement project implies that some time or money will be saved, and could represent the only case where an ROI is specifically required or generated. Although the time period chosen as a payback threshold period is subjective (or determined by internal finance/accounting policies and regulations), it generally ranges from 3 to 5 years for software projects and 5 to 10 years for hardware projects, which is the useful life of such systems. Whether or not a business case justification is required to be submitted in these categories, it may be useful to present them this way.

3.5.3 *Cost, Benefit, and Return on Investment*

Cost benefit and ROI are two of the more generic fiscal evaluations of project viability. Although the variations and level of depth that can be used in these evaluations is nearly endless, the basic approach is as follows:

- What will the new system cost, both initially and over a fixed time?
- What will the benefit of the new system be over a fixed time?
- What is the cost of operating the current system over a fixed time?

If, over a set period of time, the operating cost efficiencies of the new system, less the project cost of the new system, are greater than the cost of operating the existing system, the program, or project, is positive and should likely go forward.

The concept of ROI is similar to cost and benefit in that it looks at cost and benefit, although it also determines how much and at what point in time the project pays itself back, and will traditionally also calculate the internal rate of return (i.e., the percentage return on the investment made in the project). The conventional question here being, can the same amount of money be invested externally at a higher return?

Although calculated, cost and benefit and ROI can be highly subjective. This subjectivity lies in determining the current and future costs of the business process. Given the subjectivity of many of the cost variables involved, this can be a difficult determination to tie down. It is suggested that the utility's finance or accounting department set the rules of engagement for calculating cost and benefit or ROI. It is also suggested that they conduct the analysis as this provides an independent third-party assessment with actual money in mind vs project interests.

Savings claimed on ROI for IT projects are often realized in terms of labor hours and dollars. What this means is that the program plan is proposing that program will result in doing something for fewer hours or dollars than it currently takes. This may mean that some segment of the business is willing to either give up, or redistribute, these to-be excess resources when the effort is done. The program manager should be sure that whoever they are doing this work for is willing to state that they could be more efficient than they are now and that they are willing to meet these new efficiencies when the project is done. In the end, where a cost and benefit or ROI are required, they should be positive. If they are not, appropriate program elements should be reevaluated.

3.5.4 *Documentation*

As a program, or project, manager is responsible for the expenditure of a significant amount of money and the delivery of an equal if not greater amount of value, it is recommended that good records are kept for both planning purposes and the potential audit. As this can be a tremendous burden, a rational approach to documentation is to document only those things that are needed to ensure project success, reduce project risk, and record major decisions and accomplishments.

Because the program plan is a collection or portfolio of projects and efforts, applying basic project planning and management practices to documentation is critical. There are several structured and formalized project management methods, discussed further in Chapter 5, most of which have a generalized and universal menu of deliverable documents from which to choose and check off when done.

4.0 DELIVERING THE FINAL PROGRAM PLAN

By considering all of the variables discussed previously, a program manager should be well equipped to lay out a logical plan. There should be a clear understanding of the objectives, drivers, constraints, and influences. Consensus on the problems, opportunities, as-is state, and requirements should have been reached, and identification of the options should be made. Agreement on the selected options and solution paths to get to a to-be state should be confirmed. Activities should be modularized and assembled into testable task-based projects or efforts. Within these projects, the deliverables, scope, task schedule, budget, and resources should be identified and planned out. All of the projects, as a whole, are then arranged by priority, urgency, and/or importance, and dependencies and availability of resources. To assist change, risks should be clearly identified and mitigated. A business case showing the value against the cost and risk should be approved. A clear and concise summary presentation should be prepared for communication and approval of the plan itself. The entire package is assembled as both a proposal and a work plan, and everyone then signs on the dotted line.

4.1 Program Plan Document

The program plan document should be kept simple. An executive summary should be provided up front, and text should be at the business discussion level and not too technical. Major studies, technical evaluations, decision documents, and the like should be referenced and provided, but as appendix items. Extensive use of illustrative graphics, charts, and tables should be used where possible. Instead of actually reading every word, the average reader will probably scan headings and figures until something interesting catches their eye. As such, it is important for the program plan managers to use headings and figures to make the points they want to get across.

Essentially, the program plan is the contract by which the program will be executed, defining what will and will not be delivered and why, when, for whom, and at what cost. The program plan could very well be an umbrella collection of individual project plans, with detailed project plans being appendices (see Section 1.2 of this chapter).

Again, there is no cookie-cutter format for each of these sections. Instead, guidelines for activities to perform and pitfalls to avoid have been provided. By using these, the process of developing each section in the program plan is a matter of derivation for each unique environment. The program plan manager should start with business objectives and go from there.

4.2 Program Plan Presentation

Presentation of the program plan may be more important than the document itself. Often, few people will take the time to read a full program plan. However, they will probably scan a presentation, and it can be shown to a captive audience, whereas a full document cannot. There are likely three major showings of this presentation: (1) to executive management, (2) to the program team and customers, and (3) to the board of directors or similar governing body. Although the program plan manager may get plenty of time to present, they may also have as little as 30 minutes for the first showing and 10 minutes for the third. This presents a significant challenge in condensing the essence of what you are doing, how much it will cost, and why it is worth doing to a few minutes. The recommendation here is to avoid technology topics as much as possible and stick to business issues, solutions, opportunities, risk mitigation, and so forth. The more projects there are in the program, the less time there is to focus on any specific one (perhaps only one slide per project).

The presentation as a whole will likely have several audiences and should probably come from one master presentation, with select slides at select levels of detail shown to each group. This approach is particularly useful at executive and board levels. Although presentations should be kept simple, if specific questions are asked backup slides should be used to answer them. A "storyline" approach is recommended so that the audience can take away an understanding of the background, drivers, objectives, and solution in simple, repeatable terms.

5.0 CAUSAL SUCCESS AND FAILURE STATISTICS

The industry record for IT project success and failure rates is not entirely positive; the majority of IT projects are considered challenged or failed for various reasons. Although the trend is getting better as IT program, project management, and technical disciplines mature, there is still significant room for improvement. In preparing a program plan, it is important to apply program energies where risk mitigation and accomplishments will be best served.

Information technology program success can be a difficult to obtain. According to the 2009 *CHAOS Summary Report* by Standish Group International, Inc. (Boston, Massachusetts), 32% of all projects succeeded, which meant they were delivered on time and on budget, with the required features and functions; 44% were challenged, which meant they were late, over budget, and/or with less than the required features and functions; and 24% failed, which meant they were cancelled before completion or delivered and never used.

To be successful, it is important to make sure that these specific factors are addressed in a program plan and in actions. Although the following aspects are the leading cause of success, conversely, their absence is also the leading cause of failure (listed in order of importance) (Standish Group International, Inc., 2001):

- *Executive support*—Make sure your efforts are driven, if not heavily supported, from the top down. A project without executive support will have a difficult time obtaining funding and resources. This support may also be needed during trouble spots.
- *User involvement*—to capture and deliver the needs of users, users need to be involved. Lack of involvement may result in project rejection, whereas involvement will foster user ownership, direction, and decision making.
- *Experienced project manager*—the vast majority of successful IT projects have an experienced project manager.
- *Clear business objectives*—all parties involved in the program need to know the goals. Any work not dedicated to meeting business objectives may be a waste of time and money, placing the program at risk.
- *Minimized scope*—keep it simple. Minimize scope, schedule, costs, and risk, and increase chances of success. Break the program and projects into smaller stand-alone accomplishments, if possible.

By addressing these key aspects and those discussed earlier in this chapter, a program manager can help ensure the successful development and execution of a program plan.

6.0 REFERENCES

Brueck, T.; Rettie, M.; Rousso, M. (1997) *The Utility Business Architecture: Designing for Change;* Project No. 165; Water Research Foundation: Denver, Colorado.

Gleicher, D.; Beckhard, R. (1969) *Organization Development: Strategies and Models;* Addison-Wesley: Reading, Massachusetts.

Marston, W. (1928) *DiSC®: Dominance, Influence, Steadiness, Conscientiousness;* Inscape Publishing: Minneapolis, Minnesota.

Standish Group International, Inc. (2001) *Extreme CHAOS.* http://www.small-footprint.com/Portals/0/StandishGroupExtremeChaos2001.pdf (accessed Aug 2009).

Zachman, J. A. (2006) *Enterprise Architecture* Home Page. http://www.zachman-international.com/index.php/home-article (accessed Dec 2009).

7.0 SUGGESTED READINGS

Fisher, R.; Ury, W. L. (1991) *Getting to Yes: Negotiating Agreement Without Giving In;* Houghton-Mifflin: Boston, Massachusetts.

Kotter, J. P. (2008) *A Sense of Urgency;* Harvard Business School Press: Watertown, Massachusetts.

Schrage, M. (2000) *Serious Play, How the World's Best Companies Simulate to Innovate;* Harvard Business School Press: Watertown, Massachusetts.

Tufte, E. R. (2001) *Envisioning Information;* Graphics Press: Cheshire, Connecticut.

Chapter 5

Information Technology Capital Project Management

(continued)

1.0 INTRODUCTION TO PROJECT MANAGEMENT

Implementations of complex initiatives in all fields of human endeavor share a common set of pathways to failure that can foil the best-laid plans. These pathways to failure arise in several dimensions and require active intervention to keep an initiative on track to successful implementation. These are discussed here and also listed at the end of this chapter as a "project management checklist" for reference.

Complex initiatives range from activities such as constructing a campfire, organizing a hunt, raising a family, running a council, negotiating a trade, and building a road to waging a war, researching disease prevention, providing community water supply, engineering a system, managing information, and so forth. Complex initiatives are various forms of projects or programs. Programs are simply logical groups of projects.

Managing projects and programs has become a specialized science structured around the approximately 12 dimensions that can affect success. Some of these dimensions are obvious (e.g., time, scope, and money). Other dimensions might be less obvious and require some thought to identify (e.g., quality, resources, communication, change, and risk). Further drill-down might reveal more project management dimensions (e.g., data, procurement, and standards). The final dimension of project management that assiduous practitioners watch is integration, which enfolds and wraps the discipline of project management into its own boundary, keeping it separate from the actual execution of work under the scope of the project.

These are the dimensions of project management. They represent the levers of control to guide complex human endeavors from concept to implementation, and are also important for success. There is a common perception that information technology (IT) projects are always late and over budget. In some instances, the problem is poor management of expectations and the perception is a result of poor communication management. In others, the problem is poor management in several dimensions and the perception is the reality. Good project management keeps the project on a course to successful implementation and allows the project manager to identify and avoid the pathways to failure in multiple dimensions.

There are instances when sophisticated project management is not required for an IT implementation, such as when a simple commercial-off-the-shelf (COTS) system purchase will satisfy user needs. In such instances, business issues dominate rather than classic project management. Attention would then focus on analysis of buy vs build, negotiating for a good price, and installing the software. Standard office productivity software would likely be in this category. This chapter will not deal further with this simpler category of IT implementation; rather, it will focus mainly on project management for complex, custom development of IT systems for water utilities.

1.1 Business Challenge

All well-run businesses, including well-run water utilities, operate in a context of needing to maximize effectiveness and efficiency. Effectiveness implies getting a job done; efficiency implies getting a job done under competition for resources.

The business challenge for a water utility is exposed in a typical vision and mission statement outlining the statement, "who we are and what we do." An example might be the following: "We are your partner in ensuring a healthy water environment; we provide reliable, drinking-quality water to you and receive and treat your wastewater properly before returning it to the environment, all at an affordable cost to stakeholders."

The first challenge is in being effective (i.e., "we provide the product and service"). The second challenge is in being efficient (i.e., "at an affordable cost"). When a water utility engages in a project, the utility needs to achieve the desired project outcomes and do so at a reasonable cost. This chapter examines the many pathways to failure/success in project management of IT capital projects and attempts to illustrate that project management is both simple and difficult.

The business challenge is to execute the exact project scope within time and money constraints. These parameters are at the core of every project.

1.2 Project Management Challenge

The importance of the project management function is often underestimated. For example, people see the "oh, how simple" part readily or they may not see the need for orchestration of the many activities of the project. Indeed, the thinking may be as follows: "After all, we could just let the team of experts do their thing and the result should just naturally happen." One project management challenge is having project sponsors who think that rigorous project management is not really necessary for success.

A second project management challenge is having the technical implementation team members think that project management is not really needed. Information technology specialists, scientists, and senior management can individually see and discharge their technical scope functions, but might be unable to see that there is need for a global organizational framework around a project or that they need to defer to such a framework. Therefore, they might also conclude that rigorous project management is not really necessary for success.

A third project management challenge, and a fatal flaw for a project, is in not budgeting for the project management function at an adequate level to ensure that all the dimensions of project management are applied throughout the project's life cycle.

The project management challenge can thus be summarized in terms of sponsorship for project management, team building around an explicit project management structure, and adequately resourcing the project management function.

1.3 Information Technology Design Challenges

Information technology is a relatively new science compared to other sciences such as conventional engineering. It is a rapidly expanding field as humans discover new ways of acquiring, storing, retrieving, managing, and manipulating data and information. Each of the sub-technologies of IT is a science in its own right, and practitioners are continually specializing into narrower aliquots of their fields as they drill deeper for learning and develop applications of the new knowledge. Therefore, it is understandable that IT practitioners are generally more focused on the technology itself than on the surrounding issues of a project. Information technology has more evolution ahead as it resolves into the science and the application of IT.

An IT project often seeks to apply disparate technologies into a single objective; for instance, a new software application might seek to combine word processing and spreadsheets into a single report. Such attempts at combination complicate interconnections between sub-technologies because the word processing technology and the spreadsheet technology might originate from different firms. Even when the combination is successful, we might wonder if the resulting report mainly has the characteristics of "firm A's" or "firm B's" software. Or, is it now a "child" of this union and unique in its own right? How should the developer of this child software communicate to the outside world of users, maintainers, and so forth on what this newly spawned software can do? For example, a simple IT environment serving a small water utility might consist of three systems: a financial management system, a supervisory control and data acquisition (SCADA) system, and a maintenance management system. These could be purchased from separate vendors and could exist on separate platforms. An evolution of the system toward integration and information sharing could result in a new hybrid platform as a child of the separate systems.

With rapid proliferation of technology approaches, development of special jargon, early-to-market applications, and so forth, design of an IT project poses unique challenges. Someone has to understand the IT concepts and also understand project management concepts and bridge the communication gaps between sponsor, technical team, and other stakeholders, notably users. It seems that the universe of IT knowledge is expanding far faster than bridge builders are entering the field to support this expanding universe. This can lead to the following two undesirable outcomes: proceed without proper project management or defer the project.

A particularly important design challenge for an IT project is in bridging the communication gap between users and designers (i.e., programmers or code writers). Users know the underlying business processes and are best positioned to understand current and future business needs. However, users are generally not knowledgeable about how business processes are supported by IT itself. This is analogous to the pilot of an airplane understanding the aerodynamics of flight and navigation, but not having detailed knowledge of exactly what is happening when the throttle or power lever is moved and the engine is required to increase or decrease power output. Users are pilots; IT designers are engineers. Users/pilots know what performance envelope is required; IT designers/engineers develop the underlying machine to achieve the required performance envelope. The challenge is to find a common language for them so that expectations can be set, met, and demonstrated before the system goes live. Such a common language is largely nonexistent and, as a result, they must rely heavily on interpreters and performance testing for confirmation that at project initiation expectations *will be* satisfied and that at project completion expectations *have been* satisfied. This will be discussed in Section 2.2 of this chapter.

1.4 Obsolescence Challenge

Rapidly evolving systems generate obsolescence. Practically all IT systems implemented by a utility will require ongoing upgrade through their economic lives to remain useful within an interconnected architecture. It is not sufficient that a technology subsystem continues to function well in its core duty alone; it must increasingly communicate with newer technology subsystems in ways that could not have been anticipated at an earlier state of knowledge. There are two major ways for a program to become obsolete before the end of its anticipated economic life cycle: (1) changes in the required functionality or (2) the emerging need to integrate with other computer systems or components.

For example, a SCADA system might continue to operate a pump satisfactorily for decades after initial implementation, but the system might not be able to pass on information about how many running hours the same pump is accumulating unless the interconnectivity is compatible. If the SCADA system was originally designed only to store run hours in its proprietary time-series database for display on the SCADA screen, then that would be the limit of its capabilities. The program manager should consider that water utility business processes might have since evolved to integrate such data into other systems, such as asset management systems, enterprise resource planning systems, and so forth, and that the new connectivity requirement is to pass on information for a thousand operating assets and to do so wirelessly, at a high rate over long distances. As such, the SCADA system might now be obsolete.

All IT projects for the foreseeable future should be approached with the concept of ongoing requirements for upgrades. This has implications for management of the system development life cycle (SDLC). Project managers should be aware that time is really of the essence with these projects and that time management has two imperatives from an SDLC viewpoint. First, the program manager should capsule user requirements and expeditiously move the project to completion with minimal changes and enhancements along the way; second, the program manager should prepare the way for commencing a subsequent SDLC immediately upon completion of the current one to deal with desired changes and enhancements that came to light during the current SDLC.

Sponsors, stakeholders, and users are generally unaware that time management is so critical in IT projects compared to conventional engineering projects. This is a major mode of failure for IT projects with their high obsolescence challenge.

2.0 OVERVIEW OF PROJECT MANAGEMENT METHODOLOGIES FOR INFORMATION TECHNOLOGY

There are many ways to create a great soup using varying recipes and varying ingredients. Similarly, there are many ways of managing projects for great results using varying methodologies and resources. Some project management methodologies have been borrowed from the practice of conventional engineering and applied to IT projects with reasonable success. Some methodologies have evolved from business process engineering and can also be successfully applied to IT projects.

Good project management methodologies have common elements: structure, procedural clarity, input and reporting points for sponsor control, documentation of process to enable learning, flexibility to manage the unexpected, and so forth. In this chapter, the authors will refer to certain "actors" relevant to project management. As such, these actors are defined in general terms, as follows (it is important to note that specific projects might use variations to these definitions to suit the vocabulary and culture of the project context):

Project manager—the person responsible for executing delivery of the project. This person is the ultimate authority governing activities of the project team. The project manager reports to the sponsor.

Project team—the persons who, individually and collectively, have responsibilities to deliver each task in the project work breakdown structure.

Project sponsor (sometimes called the *"project champion"*)—the individual who takes organizational and political responsibility for achieving project goals and enables the necessary budgetary resources to sustain the project. The sponsor is often the key to managing external stakeholder relationships that are beyond the reach of the project manager.

Client—the person or organization receiving the benefit of the completed project and often the source of funding for the project.

Users—the individuals who will accept the project on behalf of the client and operate the system after delivery.

Stakeholder—any person or organization who has a valid interest in the project.

2.1 Program versus Project Management

In this chapter, the terms *program* and *project* are used interchangeably. *Program* will generally be used as a collective term for a cluster of projects, in the same way that flock is a collective term for a cluster of sheep. Although other meanings exist for the term, such other uses will be clarified in the context of this chapter.

It is often convenient and desirable to cluster individual projects into a single initiative (i.e., program) within a business environment for reasons varying from commonality of objectives, obtaining economies of scale, serving multiple stakeholder requirements, and so forth. Hence, it is quite common to see IT capital programs made up of several projects.

In the context of the practice of project management, the term *project* is also used as a collective term by the Project Management Institute (PMI) (Newtown, Pennsylvania), a widely recognized professional organization that has created ***A Guide to the Project Management Body of Knowledge (PMBOK® Guide)***, which represents a body of knowledge that codifies the discipline of project management. This discipline applies to individual projects and, equally, to programs.

2.2 System Development Life-Cycle Model

Project life cycles typically follow the classic "S" curve found throughout nature, describing such processes as transmission of infections in a population, dissemination of information into a public, cumulative work effort over a project, and so forth. These phenomena typically have a starting point and typically taper to an endpoint.

In the SDLC "S" curve in Figure 5.1, the approximate end of each phase is shown by the arrows marking definition, design, development, and release. These phases are described in more detail later in this chapter.

The SDLC model for IT projects is best portrayed as a series of life-cycle curves, as illustrated in Figure 5.2, to emphasize that *time* is of the essence. Those desirable changes and enhancements identified during the current development are typically best deferred to a subsequent cycle (or "release" or "version") to maintain focus on the initial user requirements that drove the project at kickoff. This race to completion is primarily driven by the risk of obsolescence discussed previously, although there are collateral benefits to maintaining a pragmatically narrow focus relating to managing scope creep, resourceusage, efficiency, and so forth. The SDLC for IT projects has clear phases as described in the following sections.

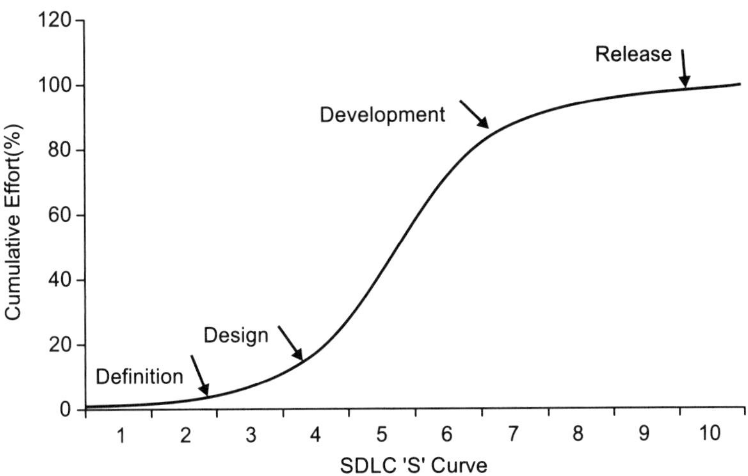

FIGURE 5.1 System development life-cycle "S" curve.

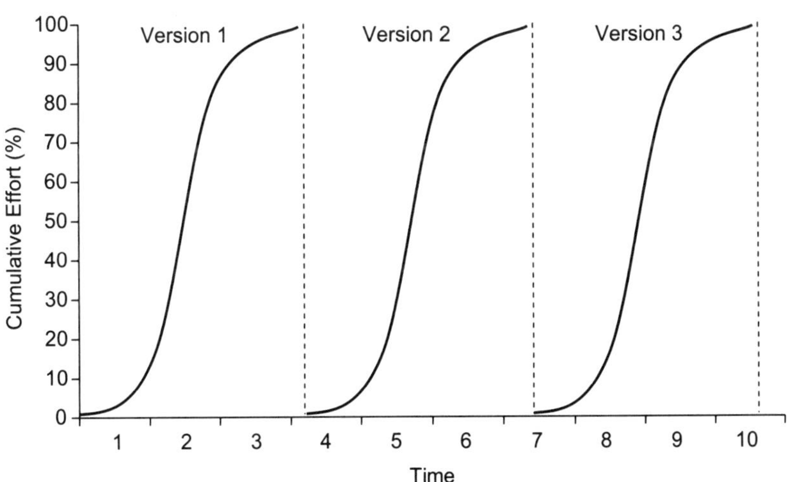

FIGURE 5.2 System development life-cycle "S" curve (with upgrades and enhancements subsequent to version 1).

2.2.1 *Definition*

During the definition phase, the project is managed through certain key activities and milestones. These include identification of need; preliminary study to frame the scope, schedule, and budget; identification of resources; empowerment of a sponsor; and authorization to undertake the project. This phase typically ends at about the time the project sponsor issues a kickoff directive.

2.2.2 *Design*

During the design phase, the project assumes more structure and involves increasingly larger numbers of people. Often, operational staff in a water utility will turn over the project to an internal or external project manager to build and maintain the necessary project management infrastructure. Activities and milestones of this phase include the following:

- Set up a project management office (PMO),
- Prepare user requirements,
- Prepare a predesign report, and
- Report to stakeholders and sponsor to receive a go/no go decision before proceeding to the next phase.

Each of these activities has associated detailed tasks, such as setting up the 12 dimensions of management inside the PMO, convening user workshops to extract and document user requirements in plain language, procurement of consulting resources to supplement in-house resources where necessary, and so forth. This phase typically ends at about the time that the predesign report has been signed off by users and accepted by the PMO.

2.2.3 *Development*

During the development phase, the project builds momentum exponentially with time. The development phase is the most intense, costly, and complex phase of an IT project. Development typically commences with expansion of the predesign report to a detailed design brief, during which several important deliverables are generated. An early deliverable is conversion of the user requirements into technical language, thereafter known as *user specifications*. Another is further drill-down of the work breakdown structure by the key technical leader (i.e., system architect for an IT project), who breaks the project into components that can be assigned to individual designers. Another is documentation of the testing protocols and plans for unit testing of the small components as they are developed, followed by integration testing of the components as they are fitted together. An important phase-end deliverable is acceptance testing by the users before allowing the new system to go live.

At each major milestone of the development stage, appropriate reporting is provided by the PMO to the sponsor and other stakeholders to ensure ongoing support is maintained through the inevitable challenges of dealing with unknowns as they arise, an unfortunate characteristic of all projects.

Throughout the project, the PMO's mandate is to continually manage the potential pathways to failure discussed previously. The reader should recall that these are in the dimensions of time, scope, money, quality, resources, communication, change, risk, data, procurement, standards, and integration. Integration here means the overlap of the other dimensions of project management, including tradeoffs and balance among combinations such as time and money, money and risk, change and risk, and so forth.

2.2.4 Release

During the release stage, system design has been tested and accepted by the user group and is ready to be used as a production system. Project management functions at this stage primarily focus on time, scope, and money. The technical resources, meanwhile, focus more intently on the schedule and cutover to an operational system, with rapid response to bugs that might not have been caught in testing before going live. The technical team leads the project through release and the PMO monitors and documents progress while maintaining the project support infrastructure, such as procurement management of any contracted resources. By this stage, assuming good project management in prior stages, the major work is largely done, budgets are mostly expended, and wrapping up the project consumes the PMO's attention.

2.3 Why Structure Is Necessary

The complexity of many IT projects can be enormous and often cannot be fully understood by stakeholders without a complete model of the project being prepared and documented through such tools as Gantt and program evaluation and review technique charts, work breakdown structures in nested hierarchies, resource assignments, and so forth. Just managing data in a major project might require development of a document management system, which, in itself, is a project within a project.

Coupled with this technical complexity is the apparently even more puzzling complexity of human interactions with personality and emotional overtones that can make a team underperform. Some of these interactions and relationships are portrayed in Figure 5.3, which shows a model of project management centered on a PMO and bi-directionally focused on external stakeholders and the project team, respectively.

Finally, it is important to consider the business context within which the project resides. The business itself might be under stress that is either related or unrelated to the project; moreover, constant competition for resources requires a sponsor to guard the project throughout its life cycle. The project manager needs the sponsor's continuous support to ensure that circumstances beyond the project manager's authority are placed before the sponsor for attention and action.

A properly orchestrated PMO allows each of the elements of the project to play out and guides the project to ensure that it stays on track despite a constant tendency, common to all projects, toward instability and failure. Without the structure of such a PMO, a complex project has little chance of meeting all of its objectives.

For practical examples of project plan structures for IT projects, the reader is encouraged to seek out readily available resources on the Internet using any of the popular search engines.

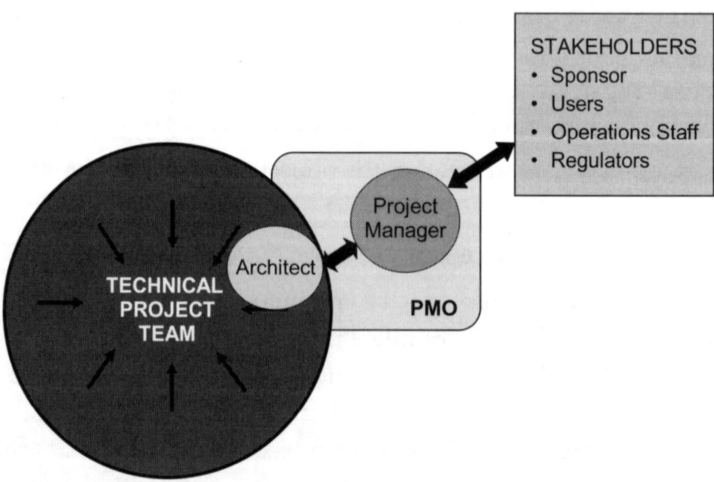

FIGURE 5.3 Project management model.

2.4 Zachman Framework

The Zachman framework is based on process analysis and business process engineering. It looks at a business in terms of layers of infrastructure and interactions of actors. It is an effective way of carving up the business into analytical elements for IT projects. Unfortunately, it is rarely well understood by external stakeholders, such as users or sponsors, and, therefore, is limited as a means of communicating with the external world.

As a means of communicating within the technical team, however, the Zachman methodology is effective. It captures and documents all elements of the project from project rationale through business needs, use cases, actor definition, modular development and interactions, object development, coding, and so forth. Zachman analysis readily lends itself to support by enterprise architectural software such as Enterprise Architect (Sparx Systems Pty. Ltd.; Creswick, Victoria, Australia) and Rational Rose (IBM Corporation; Costa Mesa, California), and so forth. These are sophisticated, valuable, internally focused tools that can be used by the technical project leader to orchestrate the internal workings of a project and provide documentation to support conventional PMI-style project management for external stakeholders.

2.5 Other Methodologies

There are other ways to manage projects that might be appropriate in certain circumstances. However, the PMI-based methodology described previously, and variations around it, constitute mainstream ideas for project management of IT projects for water utilities. Specialized methodologies are beyond the scope of this chapter.

3.0 GUIDELINES FOR MANAGING INFORMATION TECHNOLOGY PROJECTS IN WATER AND WASTEWATER UTILITIES

3.1 Understanding Multi-rational Organizations

Multi-rational organizations are ones in which several stakeholders or groups of stakeholders have divergent opinions on what is the best way to solve a problem. Generally, these opinions are valid when seen from an individual perspective; however, some or all of them might be inadequate to truly solve the problem. The project manager acting in a multi-rational organization is challenged to develop solutions within a context of strong-willed stakeholders whose buy-in and participation in a final solution is vital for success.

There are numerous examples of multi-rationality in the political arena (e.g., the European Community, the senates and parliaments of democracies, and so forth). Multi-rationality also abounds in corporate entities with consensual cultures; water utilities can be among such entities.

As an exercise in understanding multi-rationality, consider the project management challenge in supporting the technical team as a workshop convened to define user requirements. Assume that the user community is made up of a water operations group and a corporate financial management group (add more groups to fuel the fire in this analysis). Further assume that the culture of the water utility has recently undergone reorganization and that members of the two groups hardly knew each other before the project being launched.

Now attempt to develop a common set of user requirements for a new system, such as an asset management system, a SCADA system, or another system. Each subgroup of this newly convened user group would have come to the table with a set of notions influenced by the historical behaviors of the subgroup, its cultural mores, its asset management structure, its financial and accounting structure, and so forth. Each subgroup's position on user requirements would be completely valid from its own perspective, but might not be valid in support of the terms of reference of the new project.

To make progress within a multi-rational organization toward developing a common output such as a common set of user requirements, it is necessary to build a team that looks beyond the initial positions and toward the common goal. To build such a team, much preliminary work would need to be undertaken to validate the initial positions and to provide recognition of the gaps between the initial positions and the desired new project goals. Such is the nature of multi-rationality.

3.1.1 Managing Sponsors

Managing the sponsor is directed toward one purpose: communicating clearly to allow informed decision making by the sponsor where these are beyond the authority level of the project manager. A sponsor must always exist for a project to succeed. Some sponsors prefer to remain anonymous for political or other reasons and insert a proxy sponsor to front a project. Although this can be effective, it makes the project manager's work more difficult because clear communication is less likely to occur.

It is suggested that program managers define *sponsor* as that individual or group of individuals who have high-level risk exposure from the failure of the project and who have the authority to initiate the project and to provide resources for its execution. The first element is important; if a sponsor cannot be found who cares about project success and who suffers potential risks from its failure, the project is likely to fail because all projects inevitably hit rocky spots where a sponsor will be required to clear the way for the project manager.

For example, "Project A" has been initiated in a department of the utility and is midway through implementation when a challenge arises from the head of another department of the utility who would like to sponsor "Project B" to achieve the same objectives, but with a different political slant. If Project B gains momentum, the manager of Project A would have little ability to resolve which project should prevail. The sponsor of Project A would need to coordinate with the sponsor of Project B to reach consensus on how best to proceed. The project manager's responsibility for Project A would be to bring to the sponsor's attention to issues relating to the project that are beyond the authority level of project management and to seek continuing support and sponsorship for the project.

3.1.2 *Managing Users*

Users are the closest outside stakeholders with whom the project team has relationships with. They are considered *outside stakeholders* from the project manager's point of view because they are responsible for outlining functional requirements and for accepting the final deliverables produced by the project team. Users set expectations for the performance of the product at project initiation and are important as acceptance "gatekeepers" for the product nearing project end.

The project manager's relationship with the user group requires careful attention. Because users are not typically "technical" in the IT design sense, they may not be able to specify in technical terms how they want the product to perform, although they do know their requirements. Thus, the project manager must ensure that there is an accurate interpretative stage when user requirements are translated from plain language into technical language.

More importantly, the project manager must ensure that the user group and the technical team understand the function of the user group as gatekeepers for acceptance. Early achievement of this understanding will set the stage for avoiding several pitfalls relating to user acceptance. Examples of pitfalls include failure of users to set comprehensive requirements up front, failure to validate user requirements and achieve sign off for the project scope (no more and no less), failure to interpret user requirements accurately, failure to design according to the real requirements, failure to enforce acceptance of properly designed solutions if users change their minds as to scope, and so forth.

The project management office needs to explain to users and the technical team the potential failure modes and ensure that all parties understand the important roles they play in progressing toward a successful project.

3.1.3 *Managing Technical Experts*

In terms of managing technical experts, the important project management skills lie in the area of resource management. Technical issues are the domain of the technical project lead or architect and the PMO should avoid these issues. However, technical personnel in general have less tolerance for team relationship issues and can, therefore, trend toward low performance when there is real or potential discord on selecting appropriate technical solutions. A team-development model ("forming, storming, 'norming,' performing") initiated by Tuckman (1965) is a useful construct for these situations, and is briefly described here.

Experience has shown that, in many situations, formal team-building workshops pay significant dividends in trending toward high-performance teams. Left to themselves, teams tend to go through four common phases: forming, storming, norming, and performing. These phases represent successive stages in relationship building. "Forming" is stage one and occurs when two or more people assemble for the purpose of undertaking a common goal. "Storming" is stage two and occurs as people in the group discover irritations among themselves (i.e., bickering) that distract them from working on the tasks necessary to achieve the common goal. "Norming" is stage three and occurs as participants in the group start to understand and negotiate their dissonance into normalized acceptance of irritations or agreed changes to behaviors. "Performing" is stage four and occurs after the team is effectively normalized and participants then return to concentrating on executing the tasks required for achieving the common goal. The key here is that although there is no shortcut or way of bypassing the intermediate stages, with facilitation the team reaches the performing stage sooner than without facilitation.

Without the investment in team-building workshops, teams are left to transition progressively through these phases on their own. However, there is a possibility that a team could become stuck in the storming phase and fail to normalize and graduate into the performing phase.

An experienced project manager must be vigilant in monitoring so that the technical team achieves high performance and must also intervene when it is appropriate to do so when teams are not working well.

3.1.4 *Managing Vendors*

Managing vendors lies in the project management skill set of procurement management. Vendors in IT projects include individual contractors, personnel agencies, suppliers of shrink-wrapped software, hardware suppliers, system integration developers, and so forth.

Procurement management includes ensuring that appropriate contracting and technical documentation is used, quoted prices are managed, invoices are promptly paid, and so forth. A purchasing department within the utility often provides valuable support to the PMO in this area, although, occasionally, the project requirements for a major program might demand quicker delivery than a major purchasing department can deliver. In such cases, the PMO could request delegated authority through the sponsor for alternative arrangements that would satisfy policy and audit requirements while speeding up the procurement function.

4.0 STRUCTURE OF INFORMATION TECHNOLOGY PROJECTS

4.1 Define, Design, Develop, and Release Phases

As discussed previously in this chapter, define, design, develop, and release phases are logical segments for a work breakdown structure; they also represent logical go/no-go decision points for sponsor reaffirmation of need for a project. At each stage, more information becomes available in the key areas of time, scope, and money that are interesting to the sponsor. As information relevant to the time, scope, and money "basket" evolves, the sponsor has the opportunity to assess the project strategically and can guide it for best fit to strategic needs or cancel the project if future outlook warrants.

These four phases have individual project management requirements. The define phase is concerned with broadscope views, the design phase is concerned with reducing uncertainty, the develop phase is concerned with details, and the release phase is concerned with final delivery and effect on the utility's operations.

4.2 System Development Life Cycle

The key concept conveyed by the SDLC is that IT projects are amenable to cyclical iteration; thus, they do not have to be totally comprehensive at the outset. Users can then tolerate misgivings they may have that the user requirements are not sufficiently comprehensive. Otherwise, user behavior will trend toward developing the user's requirements in too fine detail and without appropriate timeliness; in other words, it will become an interminable exercise.

The cycle is intended to generate a Pareto level of functionality—capture say 80% of the required functionality before freezing the user requirements because the other 20% might not become apparent until well after the details have been developed. The cycle allows for efficient capture of the remaining 20% during the project and putting these into a basket of desirable enhancements to be implemented in the next cycle of the SDLC.

Some users have difficulty with this concept and argue that a project that is moved into production with only $x\%$ (where $x < 100$) of the functionality identified is unacceptable. If the corporate culture prevailing in the utility cannot support an SDLC approach, then the project manager must choose another methodology and ensure that, if longer project duration results from an alternate methodology, that the stakeholders accept such an outcome. As mentioned earlier in this chapter, time is of the essence with IT projects because of the potential for obsolescence, which represents a real risk for project failure.

4.3 User Participation

User participation in an IT project typically occurs near the beginning and end of the project. Users are vital for developing requirements, understanding compromises required among conflicting requirements, agreeing to a testing and acceptance protocol, and participating in acceptance testing (i.e., asking, "Does the product satisfy the signed-off requirements?").

Once user requirements have been signed off on, users generally do not have an active role in the project until acceptance testing is required, which is near the end of the development phase. This can be problematic with a project life cycle that extends over several years because individuals that make up the user group will have changed somewhat and users might have developed different views about the requirements as time passed by. Hence, it is important to charter the user group and have a governance structure for the long term.

4.4 Testing and Acceptance

Testing is formal activity that helps satisfy the PMO's requirements under quality management. There are various types of testing that occur in the life cycle of an IT project; unit testing occurs as programmers develop code; internal acceptance testing is done by the technical lead or architect; integration testing occurs as modules are fitted together; and, finally, acceptance testing is done by the user group before going live.

Unit testing, also known as *string testing*, is performed as part of code development and is undertaken by the programmer alone. The objective of unit testing is to show that the code works and satisfies the specifications handed down by the architect.

Internal acceptance testing is done by the architect or a designate other than the programmer who wrote the code. Unlike unit testing, which is geared toward showing that the code does work, internal acceptance testing is geared toward finding conditions under which the code does *not* work; hence, the requirement for independence for the internal acceptance tester from the code designer. If the code fails internal acceptance testing, the module development is reiterated until the code passes both unit testing and internal acceptance testing. This iteration is common and should not be viewed with dismay by the PMO; rather, some allowance should be built into the project to accommodate a few reasonable iterations.

Integration testing is conducted within a special testing environment, that is, a computer system separate from the development environment and closely mimicking the final production environment. Modules are migrated from the development environment for testing in the testing environment. As modules develop, they are tested for compatibility with other modules in accordance with a testing protocol designed by the architect. If a module fails an integration test, it is returned to the development environment for reworking and is then retested. No development occurs in the integrated testing environment. After development has been completed for all modules and they have been tested to the satisfaction of the architect, the user group is invited to participate in acceptance testing.

User acceptance testing is conducted in the final testing environment, which is configured to be nearly identical to the final production environment. Indeed, in some instances, the testing environment becomes the final production environment, whereas, in other instances, a successfully tested product is migrated into the existing production environment. User acceptance is conducted with users present and follows a script that tests each functionality as identified by the users in the original signed-off requirements. This is where a project manager needs to facilitate and mediate if users deviate from the principle that acceptance implies fulfillment of the project scope (no more and no less). Sometimes, users request enhancements at this stage and balk at signing off on acceptance. The SDLC approach is useful in explaining that enhancements are possible but should be held for a second iteration of the SDLC, resulting is a new version or release of the software product.

4.5 Governance

As part of any IT project, a project manager should recommend that a governance group be chartered to persist in life beyond the project and for the duration of the economic life cycle of the product. A governance group is typically composed of people from the user group, people from IT support, and people at a management level who can identify and seek sponsorship of enhancements and upgrades.

The charter for governance is to monitor the operational characteristics of the product, identify any gaps between desired and actual performance and capabilities, and conceptualize and trigger appropriate iterations of the SDLC related to the product. Governance is a legacy of any IT project. The project manager's role is to charter the group and ensure that its charter focuses on its sustainability for the duration of the economic life cycle of the product it is chartered to govern.

4.6 Upgrades and Enhancements

Software inevitably requires upgrades and enhancements. These are beyond the scope of the PMO and, therefore, not its responsibility. The project manager's role with respect to upgrades and enhancements is to capture and document desirable ones as they are mooted by users and hand them over to the governance group upon project completion for appropriate action.

It is considered poor project management practice to incorporate scope changes without sponsor approval. Upgrades and enhancements constitute projects in their own right and can consume resources well beyond the approved funding levels for the base project. Many IT project managers have followed the siren song of upgrades and enhancements only to crash on the rocks of scope violation and budget overrun.

4.7 Version Control

Version control is a subset of data management in the PMO. Data management requires that project records be captured, archived, and retrieved. Many data records are developed iteratively or become superseded. Any effort of the project team that is based on superseded data or information is wasted effort and will require reworking, plus associated cost and time effects. Even worse, any effort based on superseded data or information that goes undetected poses significant risk and quality management issues to the PMO. Therefore, it is essential that rigorous version control be applied to all data and information.

Version control should be embedded in the documentation for software code, for technical memos, for decision documents, and so forth. In fact, version control should be applied to all products within a project so that no ambiguity will exist as to what version the reader or user is focusing on when data or information are retrieved.

5.0 PROGRAM MANAGEMENT: A PORTFOLIO OF PROJECTS

5.1 Organizational Aspects

Many organizations use standard project management practices to increase the likelihood of success for their technology initiatives. Often, however, the same organizations do not routinely undergo the process of project portfolio management (PPM). This is a methodology for analyzing and collectively managing a group of current or proposed projects based on a collection of key characteristics (Wikipedia, 2009). It is a tool that organizations use to align their IT projects and initiatives with business goals; it optimizes their collective value and measures the performance of the individual projects and the collective program of projects and initiatives. In addition, use of PPM also provides indirect benefits as a result of standardization of processes, improved resource allocation, and increased opportunity for process improvement.

5.2 Methodology

There are many tools and methodologies available to assist organizations with PPM. The key challenge is gaining sponsorship for organizational commitment to start and maintain a PPM process. Project portfolio management has two distinct components: effective governance and evaluation.

5.2.1 *Governance*

Good governance of the PPM requires a committee consisting of representatives from different departments, typically middle managers, who are familiar with the company's overall core business and short- and long-term business objectives and who understand senior management's mission and vision. Members of the committee are required to have strategic perspective and to use the company's overriding goals and objectives rather than those of their departments in making their governance decisions and tradeoffs. The governance committee should develop and maintain a list of criteria based on the organizational business needs and then use these criteria to review the merit of each project individually and in combination with other projects.

The process of developing the set of evaluation criteria is of vital importance for guiding the PPM process. Establishing evaluation criteria and supporting contextual documentation must be completed before the evaluation of projects begins. Some common evaluation criteria include the following (Baschab and Piot, 2003):

- *Strategic value*—How the project will give the company new capabilities to have a positive effect, both internally and on the company's customers and suppliers;
- *Financial value*—This includes the project cost and its Return on Investment. The return on investment (ROI) should consider both tangible benefits (e.g., cost reduction) and intangible benefits (e.g., improved flexibility);
- *Adequacy of existing system*—If an IT project is intended to upgrade an existing system, the committee must consider the importance and adequacy of that system. Is the upgrade urgent? Could it be deferred for a year?
- *Risk*—The committee should consider the risks associated with undertaking the project and the risks associated with canceling or deferring the project; and
- *Interdependency with other projects*—The sequencing of a particular project may be affected by schedule and status of other projects in the portfolio.

The aforementioned represent sample evaluation criteria. Organizations may include other criteria based on their type of work. For example, a public organization may consider the effect of projects on improving public safety, water quality, or community relationships.

5.2.2 *Evaluation*

Evaluation implies developing and implementing standard and repeatable processes for evaluating, approving, and sequencing projects and initiatives.

The governance committee evaluates all projects and initiatives considered by the organization on a periodical basis. Some organizations convene their committees on a monthly or quarterly basis or to coincide with preparation of an annual budget. For a project to be considered as a part of the portfolio, the requester must submit a project proposal. The project proposal is often in the form of a fill-in-the-blanks template that, at a minimum, includes the following:

- Project name;
- Project description: a short description of the project;

- Project data: estimated cost, schedule, required resources, and risks;
- Business case for the project: describing the business benefits of the project, that is, why the project is needed and what the expected business value is;
- Financial metrics: ROI and net present value;
- Relationship to other projects, that is, does the project complement or have a dependency on other projects that are either planned or already underway;
- Risk: effect of deferring or canceling the project; and
- Alternatives: the requester must provide at least one alternative.

The committee then reviews each project based on the criteria developed earlier. Several tools are available to assist in this process. The tools use a combination of spreadsheet tables and charts as a method of ranking projects.

The committee must discuss each proposed project. The chart in Figure 5.4 shows several projects ranked based on their strategic importance and the quality of the existing system. In addition, the circle size indicates the financial commitment for the project (i.e., cost) and could be colored to indicate its financial performance (i.e., ROI). The committee must decide, for example, if the customer relationship management

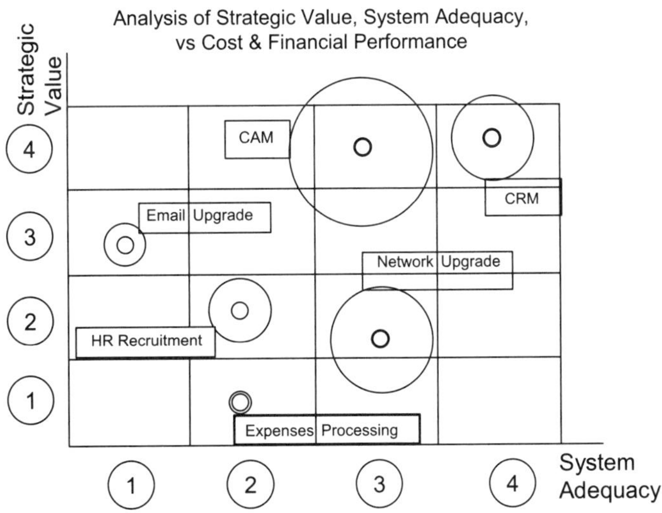

Figure 5.4 Project evaluation process—strategic importance, existing quality, cost, and ROI (Baschab and Piot, 2003).

(CRM) project should proceed, which would require a commitment of most of the company's resources. The committee may consider if there is a subset of other projects that total a similar cost as the CRM. Each project in the subset may rank lower than CRM, but, collectively, may have more benefits to the organization. In other words, the committee must decide on the set of projects that will produce optimum benefit to the organization.

In addition to determining the optimum set of projects for the organization, the group must also determine sequencing of the projects. This depends, in part, on demand management and capacity management. The IT department may not have the available resources or skills to successfully deliver all the proposed projects, and, thus, should be involved in sequencing of the projects. One of the most common reasons for project and portfolio failure is starting too many projects. Indeed, if the IT department takes on more projects than it has capacity to deliver this could lead to undesirable organizational consequences.

6.0 PROJECT MANAGEMENT CHECKLIST

The list of management areas mentioned in this section may be useful for project managers handling IT projects. The PMI *PMBOK® Guide* cited previously in this chapter recognizes and discusses many of these areas of project management. Organizing a project's document management and file structure around this checklist can prompt project personnel to pay attention to all of these areas when some areas might otherwise be overlooked. The project management checklist is as follows:

(1) *Integration management*—a PMO activity centered on coordination,
(2) *Time management*—schedule control,
(3) *Scope management*—control and containment of project goals,
(4) *Money management*—budget control,
(5) *Quality management*—quality assurance and quality control,
(6) *Resources management*—dealing with human resource issues,
(7) *Communications management*—establishing how information is shared,
(8) *Change management*—controlling scope evolution and time/money effects,
(9) *Risk management*—anticipating/avoiding project failure,
(10) *Data management*—capture of information/intellectual content,
(11) *Procurement management*—purchasing and contract issues, and
(12) *Standards management*—ensuring uniformity of design approaches.

The Project Management Institute has also developed a library of global standards under the themes of "projects," "programs," "people," "organizations," and "profession." According to PMI, the themes reflect the expansive nature of the project management profession. These standards are available for review on the company's Web site (http://www.pmi.org/Resources/Pages/Library-of-PMI-Global-Standards.aspx).

7.0 REFERENCES

Baschab, J.; Piot, J. (2003) *The Executive's Guide to Information Technology;* Wiley & Sons: Hoboken, New Jersey.

Tuckman, B. (1965) Developmental Sequence in Small Groups. *Psychol. Bull.,* **63** (6), 384–399.

Wikipedia (2009) Project Portfolio Management Web Page. http://en.wikipedia.org/wiki/Project_portfolio_management (accessed Mar 2009).

Chapter 6

Information Technology Systems—Processes and Practices

(continued)

(continued)

(continued)

1.0 INTRODUCTION

Information technology (IT) represents a significant investment; some estimates (Standish Group International, 1995) put the annual expenditure for IT projects in the United States at $250 billion. Out of those projects, 31.1% are cancelled before completion, 52.7% end up costing at least 189% of their original estimate, and only 16.2% are completed within budget and on schedule. User requirements are frequently mentioned as one of the top five reasons for project failure, and 13% of project failures are attributed to inadequate user requirements (Standish Group International, 1995). Similarly, a KPMG LLP (Toronto) survey cites that 61% of IT projects end in failure (KPMG, 1997).

Water and wastewater utilities manage infrastructure that includes facilities, plants, and pipelines; the scope and value of this infrastructure is extremely large, especially in large urban areas. As part of their traditional mission, water and wastewater utilities have many years of experience with planning and executing standard (i.e., "brick and mortar") projects for designing, constructing, operating, or maintaining this infrastructure. Critical aspects of project management for these types of projects include scope, schedule, and budget.

The scope of traditional projects that deal with design and construction of physical facilities is typically defined and formalized by design documents. Two important characteristics of these design documents are that they are typically very detailed and that the design is fairly stable, meaning that the executed projects (with small and infrequent exceptions) closely follow the design documents. Furthermore, there is typically a contractually controlled change management procedure in place that defines such events as schedule changes, material changes, and similar deviations from the detailed specification.

By the time software projects came around, most utilities had developed a strong history and tradition of project management that was based on experiences and knowledge gained from previous years of managing standard engineering infrastructure projects. Therefore, there was a tendency to apply the same methodology to the design, development, configuration, and implementation of software projects. However, lessons learned from standard civil engineering design did not translate well to software development, acquisition, and implementation.

This chapter considers modern IT system technologies, from requirements, business process engineering, and systems architecture to software development and software procurement. It illustrates the evolution and diversity of frameworks, methodologies, and practices, and creates a foundation for undertaking IT projects.

2.0 DIFFERENCES BETWEEN ENGINEERING DESIGN AND SOFTWARE USER REQUIREMENTS

In engineering projects, extensive detail in design documents is typically beneficial as it communicates more clearly what exactly the client wants and helps control execution of the project. However, in IT systems, user requirements do not necessarily follow the same logic. A classic text on software development, *Managing the Software Process*, by Watts Humphrey (1989), states

> While we sometimes get firm requirements, a software perversity law seems to dictate that the firmer the specifications are, the more likely they are to be wrong. The fact is that the demand for firm and unchanging requirements is mostly wishful thinking and may even be an unconscious attempt to avoid blame for a poor result.

This, however, seems counterintuitive. For instance, if detailed engineering designs produce better control over traditional projects, why would detailed user requirements be detrimental to the success of a software project? After all, unless user requirements are firm and detailed, how will we know that we will get what we paid for?

2.1 Managing the Unknown

The process of software development or procurement is not about implementing something that we already know or can predict with certainty; rather, it is about learning something we do not yet know and adjusting to the lessons learned. Whereas engineering design is largely fixed once completed, software implementations must be flexible enough to adjust to the emerging knowledge of the problem being solved. Ideally, software systems are implements in an iterative process, which is further discussed in Section 5.0 ("Software Development") of this chapter.

Although flexibility to accommodate new information is essential, user requirements cannot be vague and leave out important aspects of the application; they must capture the essence of the software system being developed or procured.

User requirements define the scope of the project and have a critical affect on the other two key project aspects: schedule and budget. Development of user requirements typically represents 8 to 15% of a project (Boehm, 1981), but can be traced to be the source of approximately 85% of software defects or bugs. Generally, the term *bug* describes a situation when the program does not do what the user wants it to do. The cost of fixing bugs whose source is in user requirements is typically at least one order of magnitude higher than fixing bugs where the source is in implementation (i.e., a programming bug). Defects arising from inadequate requirements occur when the software meets the needs as specified in the user requirements document, but it does not meet the users' *actual needs*. Once users start testing or adopting the system, they may "discover" needs that were not explicitly specified in the requirements; and, although user requirements are met as defined, the users may not accept the system because it does not meet their actual needs. These types of bugs can be difficult (and expensive) to address if they are discovered late in the project.

User requirements may take a different form and content depending on the nature of the project, and the requirements document will also need to be compatible with the overall project/development methodology (software development life cycle [SDLC]). Ideally, one is looking for the proper focus and level of detail to achieve optimal balance between the two following competing objectives:

- Providing enough specificity to avoid scope creep and change orders and
- Providing enough flexibility to address those issues or features that are difficult for users to specify ahead of time, although they may make a significant difference in the final user experience and system acceptance.
- Achieving optimal balance in user requirements is neither simple nor easy and, as such, "hard and fast" rules are elusive. Although the literature can provide guidelines for preparing user requirements, it is both an art and a science that especially benefits from the hands-on experience of specialists in requirements analysis (often business analysts and subject matter experts). In general, strategies for addressing the challenges of user requirements include the following:
    - Requirements may be more detailed for those projects where users are more certain about the features that they need and less detailed in areas where users have less certainty about their needs. Less precise requirements also require adjustments to the implementation plan to provide early assessment and revisions (using techniques such as iterative development and prototyping).
    - Requirements and implementation plans are different if the software will be developed or if it will be selected from existing commercial off-the-shelf (COTS) products. If selecting among available COTS products, market research can provide valuable guidance to the development process of requirements.
    - User requirements need to be integrated into the overall process (SDLC) and connected to SDLC activities such as development, configuration, and testing. The project needs to follow a clear work plan that maps the requirements into tasks, activities, resources, and schedules. Management will feel more comfortable with the project if the refinement of requirements is

planned and executed in an organized and structured manner (e.g., through iterative product reviews, workshops, and so forth).

It is also important to achieve the best balance between domain expertise and IT knowledge in developing user requirements. Without domain expertise, it is impossible to determine how the product will be addressing business processes; in addition, IT expertise is helpful because desirable functionality can be balanced with the state-of-the-art in IT.

2.2. Requirements for Software Procurement versus Software Development

In terms of software applications, water and wastewater utilities have two possible solutions:

(1) Procuring a software package from a vendor (COTS) or
(2) Developing custom software to address a specific business problem.

In either case, development of user requirements is the key first step. However, the approach to user requirements will be different for development than for procurement.

2.2.1 *User Requirements for Software Procurement*

Most water and wastewater utilities are in the public sector and, therefore, must follow a structured procurement process. Within the context of the procurement process, the goals of user requirements are as follows:

- To communicate to the vendor community the business functions that the software package will need to address,
- To provide a vehicle for the utility's selection panel to evaluate and rank the different proposed solutions, and
- To provide a basis for eventual testing and acceptance of the system.

User requirements should be developed by key stakeholders in the user community. The role of IT should be to facilitate, but not drive, the development of the requirements.

A delicate balance exists between developing a requirement/specification that is too detailed and a requirements document that omits important desired functionality. Particularly when requirements will be linked to a contract and used for eventual user acceptance of the software product, they can become very detailed and large. Such requirements can also represent a significant effort. At the Metropolitan Water District of Southern California (MWD) (Los Angeles, California), for example, the development of supervisory control and data acquisition (SCADA) requirements took about 6 man-years (three people full time for 2 years) to complete during the last major SCADA upgrade.

It is often difficult to describe the functionality of the system in a document in such a way that the user can have a clear sense of how the application will work. It is not easy for end users to articulate their tasks and the features that a software system should have. To establish the proper level of detail, it is useful to do the following:

- Identify key end users and stakeholders and conduct interviews. Find out which key features are important to them.
- Conduct a quick, preliminary assessment of the marketplace: find out what products are available and the general price range.
- Review information about this type of software product that may be available from independent IT research organizations such as at Gartner Inc. (Stamford, Connecticut) (http://www.gartner.com/) or Forrester Research, Inc. (Cambridge, Massachusetts) (http://www.forrester.com/rb/research).
- Provide end users with information about several different products, and possibly download papers with case studies or trial copies of the software to demonstrate the functionality that these products provide. It is important that users get such exposure before the selection process begins; seeing what is available on the market will help users define what they want from a system and will provide a framework for their requirements.
- If users have never used a computer system for this business process and have no experience with similar products, it is often a good idea to issue a request for information (which does not result in a contract) and ask vendors to come in and show their products in an informal, educational setting.
- Contact similar utilities to find out what products they are using and if they are happy with them.

In addition to functionality, requirements should also identify integration needs with other (i.e., legacy and future) software systems. Integration is increasingly being viewed as an important aspect of a software system's functionality. This area typically engages business process management (discussed in Section 3.0 in this chapter) and IT architecture (discussed in Section 4.0).

User requirements for software procurement must accomplish the following main objectives:

- Provide an efficient vehicle to communicate user needs to the vendors who will bid on the system,
- Define the framework for selecting the product and for making purchasing decisions, and
- Provide the basis for acceptance testing.

The user requirements process, and the requirements document, need to fit well within the organization's procurement model (e.g., request for proposals or request for qualifications processes) (see Section 6.0, "Software Procurement," in this chapter).

2.2.2 *User Requirements for Software Development*

When an organization determines that its business process and system needs require a custom application, the next important decision is whether to use internal resources (staff) to develop the software or to hire an external firm. Because hiring external resources typically involves preparing contracts, there is a greater chance that the user requirements (and the system acceptance criteria) will be well defined. "Home-grown" systems developed by enthusiastic staff often do not start from a user requirements document; rather, they are developed "spontaneously" and without a structured process. In the long run, such customized code can become a costly maintenance problem when the staff member leaves, gets reassigned, or loses interest in the software. The problem is often exacerbated by lack of proper documentation.

Regardless of who is developing the software, defining user requirements is still essential. Per a broad industry survey (Schwaber, 2006), development of a requirements document benefits from the following best practices:

- Balance business and IT involvement; the best results are obtained when there is a proper balance between the two.
- Recognize that text is not the best medium; the primary role of user requirements is to *communicate* between the users and providers of the system. This is often done more efficiently with drawings, diagrams, and sketches rather than relying on text only. "Use cases" are increasingly accepted as an efficient and effective methodology for defining user needs.

- Secure proper training for business analysts. The importance of a skilled business analyst cannot be overstated. The development of user requirements needs to constantly "bounce between" the "what" (business needs) and "how" (technology). Best results are achieved when the process is integrated and iterative.

Use cases (Armour and Miller, 2001; Cockburn, 2001) provide a methodology that allows for better communication between users and technologists. This methodology includes text-based tools and tools that include graphical representation to describe the functionality. (An example of a generic use-case template can be obtained at http://www.processimpact.com/goodies.shtml.)

Use cases were used to define the functionality of a complex modeling and decision support system at MWD. Interviews were conducted with stakeholders using the template shown in Table 6.1.

Use cases were organized in a hierarchical structure with parent–child relationships. Detailed functionality was defined in a structured, hierarchical set of use cases.

TABLE 6.1 Template for interviewing stakeholders.

Business process/ decision elements	Explanation	Suggestions/options/examples
Name	The name of the business process or software function	Regulatory reporting, Water allocation for power generation, crew scheduling, influent pumping control, etc.
Objective	The objective of this business process/software function	Establish control setpoints, dispatch repair crew, react to flooding reports, emergency response to security concerns, prepare regulatory report, prepare schedule for operator crews, perform unscheduled maintenance, perform scheduled maintenance, etc.
Description	A narrative description of this business process/software function. List the steps that are included in this business process and also list actions that are taken.	Describe the steps that are part of this business process and list actions that are included

(continued)

TABLE 6.1 *(Continued)*

Business process/ decision elements	Explanation	Suggestions/options/examples
Trigger	What initiates this business process and/or demands this functionality?	Regulatory requirement (for a report?), change in water demand, detection of SCADA alarm, storm event, capital program requires input, change in labor contracts, etc.
Timing	How often does this business process occur?	Online automatically, daily, weekly (schedule maintenance activities?), monthly (reports?), every 2 years (update to capital improvement plan?), etc.
Decision level	What type of a decision/business process is this?	Strategic, tactical, operational
Actor	Who is the primary person responsible for this business process/ decision?	Plant manager, shift supervisor, operator, process control engineer, maintenance manager, planner, other
Measures	What is the measurement used to evaluate this business process?	Revenue generated, water quality, power costs, deviation from setpoint, capital costs, compliance, system capacity, etc.
Target	The target performance measurement of this business process	Total dissolved solids level, minimum required volume of water for power generation, etc.
Parent names	If this decision/business process is part (or one "path") of other (higher-level) decisions, put the names of those higher decisions here.	Establish hierarchy of functionality
Category	What is the category of this business process?	Process control, planning, regulatory compliance, maintenance, other
Support system	Is this business process/decision currently supported by another specific software package?	Name of the product/system that is primarily used to make this decision today (e.g., Maximo asset management system, GIS, distributive control system, etc.)

The following example shows how one of the top-level use cases (Create Simulation Scenario) is broken down into more detailed ("child") use cases:

- Create_Simulation_Scenario
  - Setup_Simulation_Parameters
    - Retrieve_Defined_Sim_Par
    - Edit_Defined_Sim_Par
  - Select_Geographic_Area
    - System_Wide_Scope
    - Predefined_Subarea_Scope
    - User_Defined_Subarea_Scope
  - Define_Scenario_Demands
    - Retrieve_Measured_Demands
    - Modify_Measured_Demands
      - Manually_Modify_Measured_Demands
      - Define_Modification_Formula_Measured_Demands
    - Obtain_Future_Demands
      - Retrieve_Basic_Demand_Projections
      - Modify_Future_Demand_Projections
        - Manually_Modify_Future_Demands
        - Define_Modification_Formula_Future_Demands
      - Configure_User_Defined_Demands
  - Define_Scenario_Inflows
    - Retrieve_Measured_Inflows
    - Modify_Measured_Inflows
      - Manually_Modify_Measured_Inflows
      - Define_Modification_Formula_Measured_Inflows
    - Obtain_Future_Inflows
      - Retrieve_Inflow_Projections

Modify_Inflow_Projections

Manually_Modify_Future_Inflows

Define_Modification_Formula_Future_Inflows

Configure_User_Defined_Inflows

Define_Scenario_Network

Define_Structural_Changes

Retrieve_Predefined_Network

Edit_Predefined_Network

Define_Equipment_Availability

Define_Conduit_Availability

Setup_Withdrawal_for_Power

Retrieve_Power_Schedule

Edit_Power_Schedule

2.3 Nonfunctional User Requirements

In addition to end-user functionality, other aspects of a software application are important to consider, including

- Scalability and performance (e.g., speed of execution),
- Security,
- Reliability,
- Availability, and
- Maintainability.

These aspects of a software system will have a significant effect on the success of the system, on user acceptance, on long-term viability of the software, and on the overall cost of ownership. Therefore, these aspects must be carefully considered whether purchasing or developing a software system.

It is important to note that these aspects must be considered from the beginning of system development because they cannot be easily fixed by making modifications or adjustments to an already developed system.

2.4 Software Tools for Managing User Requirements

Several vendors offer software tools for requirements management. A partial list of these is provided in Table 6.2.

Useful information about methodology for developing user requirements can also be obtained from research organizations such as the Carnegie Mellon Software Engineering Institute (SEI) (Pittsburgh, Pennsylvania) (http://www.sei.cmu.edu) and online newsletters and resources (e.g., http://seilevel.com/resources/whitepapers.html and http://www.requirementsdevelopment.com/).

General rules for selecting user requirements tools are as follows:

- Avoid tools that are too complex,
- Select a tool that is compatible with the organization's software development methodology (SDLC), and
- Make end-user comfort the dominant selection criterion.

Before selecting a tool, it is a good idea to review broad industry reports that provide independent assessment of different vendors and tools (Schwaber and Gerush, 2008).

Table 6.2 Examples of vendors and products for managing user requirements.

Vendor	Product user requirements management system
Borland Software	CaliberRM
Compuware	Optimal Trace
Hewlett-Packard	Quality Center
IBM	Rational RequisitePro
Microsoft	Visual Studio Team Foundation System
MKS	MKS Integrity
Rally Software Development	Rally Enterprise
Ravenflow	http://www.ravenflow.com/news/pr_090606.php
Serena Software	Dimensions RM
Telelogic	DOORS

3.0 BUSINESS PROCESS MANAGEMENT

This section introduces concepts in business process management (BPM), a convenient and intuitive way to think about, organize, and automate business activities that all water and wastewater supply, treatment, and distribution organizations deal with on a daily basis. These concepts are useful to anyone responsible for ensuring efficient operations in a utility, especially, but not exclusively, those activities that can be enhanced by IT. Business process management can be used to understand the utility better without ever touching your computer, but it also provides powerful mechanisms well suited to automation of business processes. Software packages that you might buy for a specific purpose, such as those for finance, employee relations, or customer relations, are likely built on a BPM framework.

3.1 Introduction

Many people start their business day by first opening an e-mail program or Web site, reading e-mail messages and prioritizing work requests in the messages, and then begin completing each requested task. This high-level description may be represented as a simple business process model, or workflow, like that shown in Figure 6.1 or, because the tasks involve loops, a complex model like that also shown in Figure 6.1.

The workflows shown in Figure 6.1 were created with business process management software. Tools like these enable creation of computer programs that automate many tasks in a typical business. Moreover, the methods of business process management organize and give context to business activities so that they can easily be measured and controlled.

Business process management is an organization management theory and the set of methods and tools for representing, analyzing, automating, and controlling the processes that define a business or organization. It is founded on the belief that a business comprises a set of activities or processes that require resources and can be defined and organized in important, useful ways and measured and managed to achieve business goals. Business processes are managed by controlling inputs to business activities, the nature of the activity as indicated by process metrics, and the resources required to complete an activity. Business process management is strongly influenced by technology because advances in computers and software have enabled BPM capabilities that previously were difficult to achieve. Business process management is often associated with IT systems because BPM provides the essential business context for implementation of IT systems. Historically, IT motivated important developments in BPM because most IT systems were designed to support business

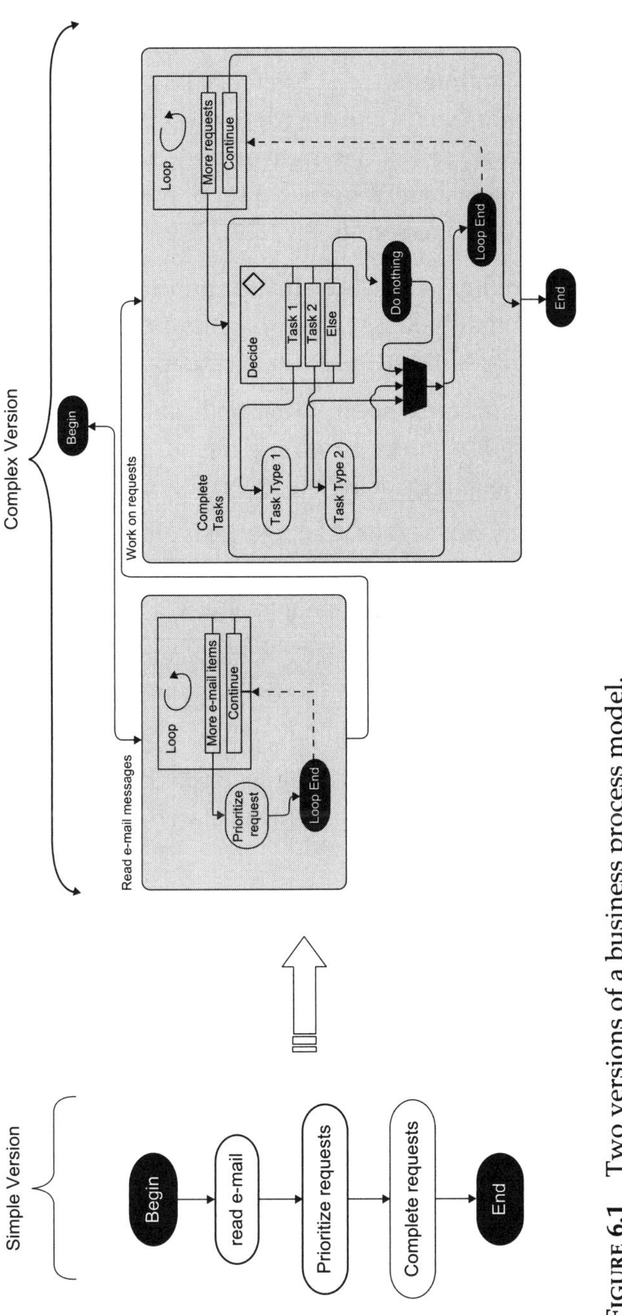

FIGURE 6.1 Two versions of a business process model.

activities. Information technology has always had a major effect on areas such as business process modeling and analysis, business process reengineering, workflow automation, simulation, user interfaces, and so forth. Business or organizational goals define the value of the organization and are the highest priority; thus, BPM provides a rationale for IT systems and drives IT system evolution.

Business process management comprises a set of methods or best practices and software tools, including combinations of the following:

- *Business process reengineering*—A discipline for business process change or transformation that became popular in the 1990s with the publication of *Reengineering the Corporation* (Hammer and Campy, 1994);
- *Business process modeling*—The set of methods and tools for constructing and analyzing models of business processes, including models developed using business process modeling notation (BPMN);
- *Business process automation*—The set of methods and tools for creating software applications that execute business process workflows, including programs that use business process execution language;
- *Business intelligence, data mining,* and *data warehouses*—The set of methods and tools for deriving information and knowledge from raw data collected from monitoring business activities;
- *Business rules management systems*—The set of methods and tools for representing rules, typically expressed as "if-then" statements, which drive business decision-making processes;
- *Enterprise application integration*—The set of methods and tools for linking or integration of software applications together to support the needs across the business locally and as a whole; and
- *Business activity monitoring*—The set of methods and tools for developing indicators or metrics for business processes for the purpose of display within graphical user interfaces.

Business process management applies to all levels within and across an organization or enterprise, as shown in Figure 6.2. Figure 6.2 depicts a hierarchy of organization functional units from base-level tactical operational processes that handle day-to-day activities, to the organizational services that set operational policies by managing operations to meet organization goals, to strategic business management focused on long-term goals.

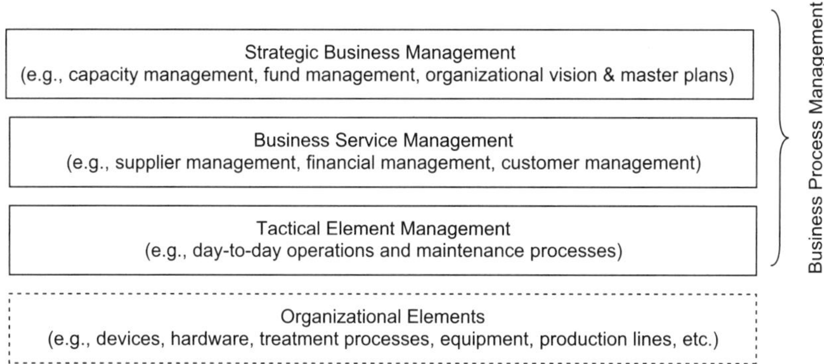

FIGURE 6.2 Hierarchy of functional units or layers in a business or organization.

Business process management's appreciation for integration across the enterprise is a key reason for its adoption. Every layer of management in an organization can benefit from BPM because processes defined within each layer represent how work is done, what information is required, who (i.e., personnel) is affected, what other processes are affected, and the relation of the process to achievement of the goals of the organization. Business process management makes these processes explicit, shows how they are interrelated, explains how each process helps create value, and provides a basis for monitoring and improvement, including rationalizing the use of IT to support a business process.

This section presents essential information on BPM, including methods and software applications and common BPM system designs and examples. First, the value of BPM in water and wastewater utilities is described, explaining motivations, benefits, and enablers that encourage adoption of BPM. Business process management concepts are described using a common maintenance management process, which demonstrates the different components of BPM suites. In addition, step-by-step methods are presented that will help the program planner at every stage, from obtaining consensus within your organization to evolutionary development and continuous improvement. Finally, a discussion of BPM trends provides additional perspective on new directions in BPM practices, tools, and methods.

3.1.1 Business Process Management Value in Water and Wastewater Utilities

The four major drivers for implementing BPM in water and wastewater utilities are to improve organizational economy, safety, flexibility, and environmental protection. The ability of BPM to explicitly define processes, expose opportunities for improvement, and support automation enhances a utility's ability to optimize around each driver.

3.1.1.1 Economic Performance

Minimum operational cost and maximum return on assets are common goals for a utility driven by the need to satisfy ratepayers. To achieve this goal, utilities take steps to improve efficiency and productivity while minimizing the probability of shutdowns or abnormal operating conditions. It is significant to note that economic performance is not always an objective for a utility because other drivers may be more important.

3.1.1.2 Safety

Safety is an essential motivation for all industries and a key element in most designs. Besides the obvious concern for the health and well-being of employees and customers, the practical reason for this is also the high cost associated with failures, injuries, and loss of life. It is estimated that abnormal conditions cost industry billions of dollars each year and result in injuries to people and damage to equipment and the environment. Safety concerns sometimes may be at odds or compete with other organizational goals, including flexibility, economy, and environmental protection objectives.

3.1.1.3 Flexibility

In utilities, the life cycles of plant assets and the "product" (e.g., treated water or wastewater, power supply, and a clean environment) must be managed. Specific management objectives depend on the time of year, the condition of equipment, and periodic or occasional events such as storms, power outages, and process upsets. In addition, customer demands may change. These factors make it necessary for utilities to design facilities that are flexible and adaptable and to maintain facilities that run efficiently under different or time-dynamic conditions.

3.1.1.4 Environmental Protection

Environmental protection is an important driver for utilities, which are viewed by the community as stewards for both significant regional assets and for the overall health of the natural environment.

3.1.1.5 Consensus Building

In addition to the relationship of BPM to organizational drivers, BPM provides a way to achieve consensus within a utility. Business models can be prepared for current or "as-is" systems and for proposed "to-be" systems. In some industries (e.g., supply-chain and telecommunications industries), business modelers have identified common patterns of business processes and developed reference models for their industries that include descriptions of best practices and standard metrics for measuring process performance. Development of standard reference models has resulted in creation of benchmarks, or quantitative baseline values, for performance metrics that others within the same industry can use to measure relative performance.

3.1.2 Business Process Management Enablers

Several factors in an organization can help program managers build a vision and consensus around BPM. These factors include IT spending plans, availability of IT technology that is practical rather than brand new, and acute requirements such as the need to make knowledge more transparent (e.g., to address succession planning) and to reduce complexity (e.g., by organizing business processes and making them accessible to everyone in the organization). Together, these factors help build a strong case for better planning of business functions and supporting IT systems.

Information technology spend plans are developed around a vision for the utility. However, a vision statement is high-level and conceptual without the detail needed to identify a stepwise process for implementing the vision. Business process management provides a mechanism for bridging the gap between conceptual design and an execution plan that explains how to get from overall intent to specific IT systems that support the business as a whole and component parts of the business. Long-term plans and associated spending are reinforced by a plan (including BPM blueprints) that describe how an expense (a) provides a certain level of capability, (b) meets performance objectives defined by clear metrics, and (c) supports the organization as a whole.

3.2 Concepts in Business Process Modeling

Business process modeling represents the operations of a business. In the sense that business models can be constructed from fundamental building blocks, just like models of water or wastewater treatment processes are constructed from chemical, biological, and physical components, business process modeling constructs from business tasks, workflows, resources, dependencies, and measurements.

The definition of business process modeling, as described in Section 3.1 of this chapter, identifies the building blocks that must be defined in modeling a business. The five important concepts are

(1) *Tasks*, or activities, combined in a flow of work or *process*;
(2) *Work objects*, including tangible (e.g., product) and intangible (e.g., service) objects;

(3) Business *rules* that represent how decisions are made;
(4) *Resources* required for a task and these can be either machine or human; and
(5) *Metrics* that define degrees of success or failure for a task.

Business process models can be (a) used as reference material (as documents for viewing and discussion), (b) simulated in computer software to examine model behavior, or (c) converted to real-time business workflow management software applications that automate the business process. This section describes each of the components of a business process model, with examples of using the models in all three ways. It is important to think of the model as distinct from its use. The model is a static description that creates benefits in many ways depending on how it is used.

3.2.1 *Tasks and Workflow*

Business activities are chunks of effort performed either by a person or machine to realize value in the business. Linking activities, or tasks, into sequential or parallel (when done by other persons or machines) chains results in a process or workflow (in this section, the terms *task* and *activity* are synonymous). Any business process can be defined this way, like the workflow introduced at the beginning of this section (see Figure 6.2).

Each box in the diagrams of Figure 6.2 represents a task, or activity, that requires effort. The workflow comprises the various sequential tasks. Parts of the workflow can be contained in groups (e.g., the "complete tasks group"), which allows for a hierarchical organization of workflow. This is illustrated in Figure 6.3, which contains another workflow with hierarchical levels shown in different windows.

Paths between activities depict sequence or ordering, which is meant to convey the important constraint that a preceding activity must be completed before a following activity begins. Where activities can occur at the same time, then a different, parallel path is created in the model.

Workflows are triggered by events. In the workflows shown in Figures 6.1 and 6.3, the "begin" and "arrow" blocks at the start of the workflows represent the event that starts the business process. In a real-time workflow scenario, the trigger could happen on a periodic or scheduled (including irregular interval) basis or when a specific request is received, perhaps by another business process.

Workflows can be simulated. Most commercial business process modeling tools provide execution of workflows in both real and simulated time. The value of simulation is the ability to create different use scenarios and examine them in fast time to verify that the model is working as designed and to validate that it achieves the

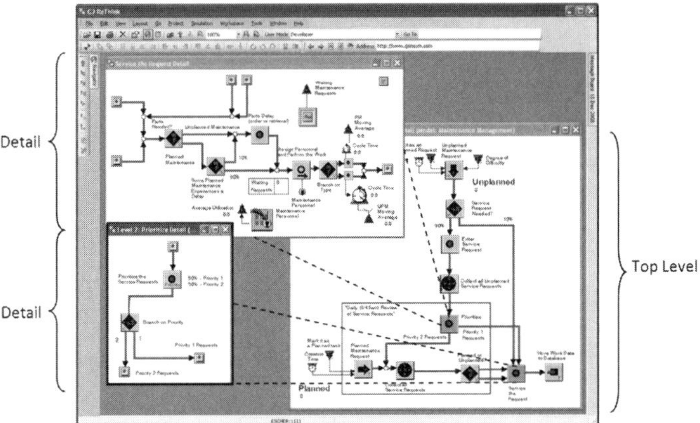

FIGURE 6.3 A hierarchically organized business process model.

intended objectives. Simulation is also useful for teaching. Real-time workflow is different in that the real-time clock cannot be accelerated and events in real time may occur at any time and must be handled accordingly.

Workflows can process everything automatically or they may require human intervention or both (a hybrid). These modes are supported by business process modeling tools. Automatic processing is common and possible when every task in the workflow has programmatic access to everything it needs to complete its work (e.g., data, resources, etc.). Examples include a business process for checking inventory stored in a database for line items in a customer order and assigning a work order to a maintenance employee based on stored service requests and worker schedules. Other business processes require human intervention to supply data or to make decisions. The inventory example could be triggered by someone entering an order on a Web page; the maintenance service request assigned to a worker may require a manager's approval.

3.2.2 *Work Objects*

Work objects, such as sets, or data, and documents, are required or produced by a business activity. Work objects are typically "information" structures that are stored in a database. For example, in a maintenance order processing activity, the order must be examined, decisions made based on order characteristics, and changes made to the order (e.g., assigned personnel). Because paths that connect activities in a business process model often start when a document or data are received, process models trigger tasks when work objects "arrive" at the task. For example, in Figure 6.1, the "begin" event triggers the "read e-mail" activity, which transforms the trigger into a "request" work object and then sends it along to the "prioritize requests" activity and so forth.

Definitions of work objects, including their name, characteristics, data they contain, image displayed (if needed) in a user interface, and so forth, are often unique to the type of business or organization. For example, utilities deal with objects like work orders, maintenance orders, or capital plans. These kinds of objects are often defined in enterprise data models developed for the organization. Creating a good enterprise data model is a significant effort that is undertaken so that the organization's objects, attributes, and relationships are well defined. This allows application architects to create clear, unambiguous business applications and rules. For this reason, it is a good practice to align the work objects used in a business process model with those objects defined in the organization's enterprise data model.

3.2.3 *Decision Tasks and Business Rules*

All business process models contain decision rules that must be defined and managed. For example, in Figure 6.1 there are rules for prioritizing tasks and rules embedded within the tasks. The term, *rule,* can be broadly defined to include any type of algorithm or procedure for making a decision, or narrowly defined as "if-then" statements. A business rule takes the following form:

If <antecedent1> **and/or** <antecedent2>...

Then <consequent1> **and** <consequent1>...

For example, a simple rule to set task priority might be

If the e-mail sender is Mr. Boss **or** the e-mail sender **is** Mrs. Wife

Then the priority of the task **is** high.

In these statements, the words in bold are reserved words found in many commercial rule languages. An <antecedent> is a phrase that can evaluate to "true" or "false" such as "the e-mail sender is Mr. Boss." A <consequent> is a phrase that can set a value for some object's attribute. Consequents typically involve an object, its attribute, and a value for the attribute. For example, "the priority of the task is high," where the object is a *task*, the attribute being set is the *priority* of the task and the priority's value is *high*. Consequents may also start or call a software application (e.g., to display a screen, perform a calculation, etc.). This simple syntax or grammar for rules enables representation of many types of decisions.

Many rule taxonomies, or types, have been developed along with methods for elucidating, organizing, and recording rules (Von Halle, 2002). Business modelers have established a convenient way to link business process models with business rules, that is, one that is both intuitive and enables rules to be organized in groups or "rule sets" so that it is not necessary to deal with the complex chaining of rules.

A convenient rule of thumb for combining business rules and workflow is to remember that decision tasks are the anchors for rules in a process flow. Here, a decision task is a kind of task that can be represented using rules in the form described previously. For example, the process depicted in Figure 6.4 shows a health insurance claim processing model that has a single decision block labeled "route claim based on physician's location." Attached to this block is a single rule, as follows:

if the PhysicianLocation of Claim = "Cincinnati"

or

the PhysicianLocation of Claim = "Boston"

or

the PhysicianLocation of Claim = "Toronto"

then

conclude that the route-to-csr of the claim_header of Claim = true

Where

csr = customer service representative.

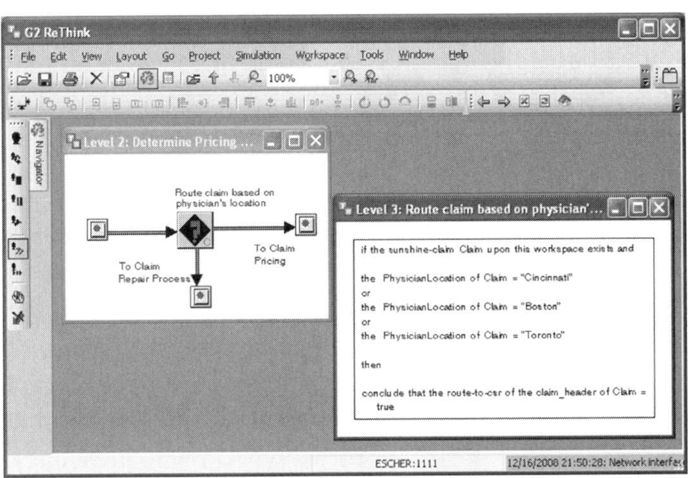

FIGURE 6.4 A decision block in a business process model.

This rule looks at the *PhysicianLocation* attribute of the *Claim* object and if it has a value of "Cincinnati," "Boston," or "Toronto," then the value of the *route-to-csr* attribute of the *claim_header* of the *Claim* is set to "true." In the business process model, this rule is executed or "invoked" when the work object (a *Claim*) enters the decision block. If any of the antecedents of this rule are found to be true, the consequent part of the rule "fires." Part of the logic of the block is to look at the *route-to-csr* attribute of the work object after this rule is invoked then route the work object to one of the block's outgoing paths depending on the *route-to-csr* value. In this model, if the rule consequent fires then the *Claim* object is routed to the "to claim repair process" task output path. The rules represent the logic that if the *Physician* shown on the *Claim* is from one of the cities named in the rule, the *Claim* should be forwarded to a customer service representative for further processing.

Rules-driven business process models comprising decision tasks like that shown in Figure 6.4 may contain hundreds of blocks and thousands of rules. Each decision block may have one or many rules associated with it. The rules may be entirely independent or the rules, through their antecedents and consequents, may link together to form a chain of inference. To manager all of this, a knowledge repository is developed to contain all the rule sets and process models, including rules and processes with different versions, owners, categories, and so forth, so that they can be rapidly accessed, modified, viewed for editing, and exchanged with others (see Section 3.3.5).

Business process models that contain if-then rules are easy to manage and easy for nonprogrammers to access, understand, modify, and extend. When process models manage real-time workflows, this accessibility allows non-IT resources such as business analysts to be involved in the creation and ongoing maintenance of the business logic that, in the end, must create value for the organization. This reduces the total cost of ownership of the workflow application.

3.2.4 *Resources*

Work activities have to be completed by some type of resource. These resources can be machines (trucks to move inventory, computers to perform calculations, treatment processes, etc.) or people (supervisors to check maintenance orders, laboratory analysts to conduct tests, managers to approve purchases, etc.). When a business process is entirely automatic, it is referred to as a *machine-centered process*. Here, *machine* refers primarily to a computer, although, in general, other types of machines may be involved in executing the business process workflow logic. This type of process requires no human intervention, except perhaps to monitor performance or handle faults that the process cannot handle itself. Other business processes are and performed entirely by people manually. For example, many manufacturers in the early part of this century handled order processing, inventory control, and production (except those manufacturers who used machines) entirely by hand. The most common modern business processes are either human-machine or fully-automated machine processes.

Resources are important to business processes because they constrain processing. Figure 6.5 presents a model for the maintenance management process that (a) assigns work orders to maintenance personnel and (b) waits until the work task assigned to selected personnel is complete. The blocks representing activities in business models are "multithreaded," that is, each block spawns a work activity every time it is triggered and each block can manage multiple activities simultaneously. In this example, the "assign personnel and perform the work" block might be invoked hundreds of times to handle many work orders.

The aforementioned activity or block at the center-right of Figure 6.5 has attached to it a small icon labeled "maintenance personnel." The purpose of this block is to check the schedules of all maintenance personnel and to select the person(s) with the required skills and availability for each work order. The "pool" of maintenance personnel shown at the bottom of the workflow in Figure 6.5 contains a database of personnel shown to the right in Figure 6.5. Each icon shown has attached to it a small clock that defines the availability of that person for work. No work can be completed without an assigned resource. In addition, limited resources may limit the overall effectiveness of this maintenance management workflow.

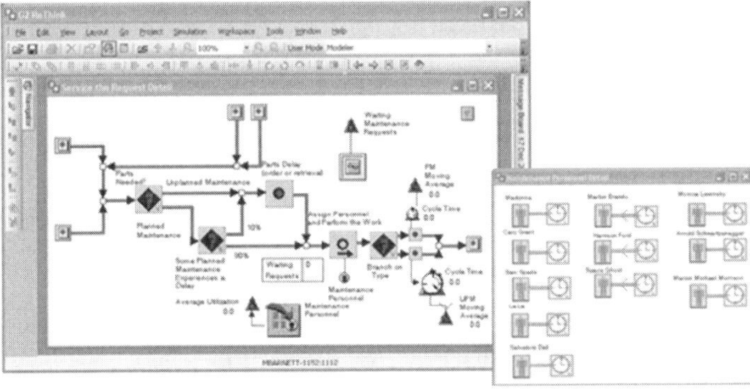

FIGURE 6.5 A business process model containing resources.

Resources such as the personnel shown in Figure 6.5 have a cost-per-time unit (the organization's cost for that resource) and a cost-per-use unit (the cost incurred each time the resource is used). In addition, the activity itself may have these same costs, reflecting the fact that some cost is incurred over time and for each time the activity is invoked. This type of information allows generation of activity-based costing information that can be used to analyze the performance of a business process from a financial perspective.

3.2.5 Metrics

An important precept of the Six Sigma (Motorola, Shaumburg, Illinois) continuous improvement methodology is that "you can't control what you can't measure." A business process model's value as a mechanism for controlling the organization is determined by the richness of the metrics that can be obtained from the model.

The three generic types of metrics that can be obtained from a business process model are (a) cycle times, (b) throughputs (or the reciprocal turnover), and (c) activity or resource costs. Cost metrics were discussed in the previous section. Cycle times are differences between the times when two important events occur, for example, the difference between when an order is placed by the customer and when the company received the cash payment from the customer, or the difference between when a maintenance work order is received and when the work is completed. Throughputs are metrics such as the rate at which work is completed, the turnover of parts inventory, and the rate at which work orders are received over a period of time.

Clearly, selecting the best metrics for a business process is important if process effectiveness and efficiency are to be continuously improved. The following are guidelines for creating effective metrics:

- Define the metrics that are currently used. At a minimum these have been proven useful in the past;
- Create cycle times around processes that take the most time and thus are likely bottlenecks. Similarly, for costs, define metrics for the activities and resources that are most expensive; and
- Define metrics (such as throughputs and turnovers) that reflect dynamic changes in the business process. Because time dynamics are significant in many business processes, it is best to maintain measures that help to understand the effects of change on the process.

3.3 Business Process Management Software Suites

Business process management's focus on process improvement generates many best practices for methods, covered in Section 3.4 of this chapter, and software, which is covered in this section. This section focuses on specific software applications that exist in a BPM system and how they are integrated to create BPM suites. Some software vendors provide individual software components described herein; however, the majority of BPM vendors offer most or all of these software components as a bundle. Sinur and Hill (2009) defined the following 10 different components to a BPM suite:

- Process execution and state management engine (infrastructure),
- Model-driven composition environment (applications),
- Document and content interaction (applications),
- User and group interaction (interfaces),
- Basic connectivity (interfaces),
- Activity monitoring and business event support (applications and infrastructure),
- Simulation and optimization (applications and infrastructure),
- Business rule management (applications and infrastructure),
- Management and administration (infrastructure), and
- Process component registry/repository (infrastructure).

These 10 components can be simplified into four groups (i.e., infrastructure, applications, interfaces, and applications and infrastructures, shown in parenthesis) as described in Section 3.3.1 of this chapter.

3.3.1 *Overview: A Typical High-Level Business Process Management System*

Figure 6.6 depicts a typical high-level BPM system comprising several software applications (center) and shows the roles of three major user groups (right) and common data sources and targets (left). Software applications in this BPM suite can be divided into interfaces, applications, and infrastructure.

Interfaces comprise all the bridges, adaptors, or connectors that provide for the transfer of information between BPM applications and other external applications,

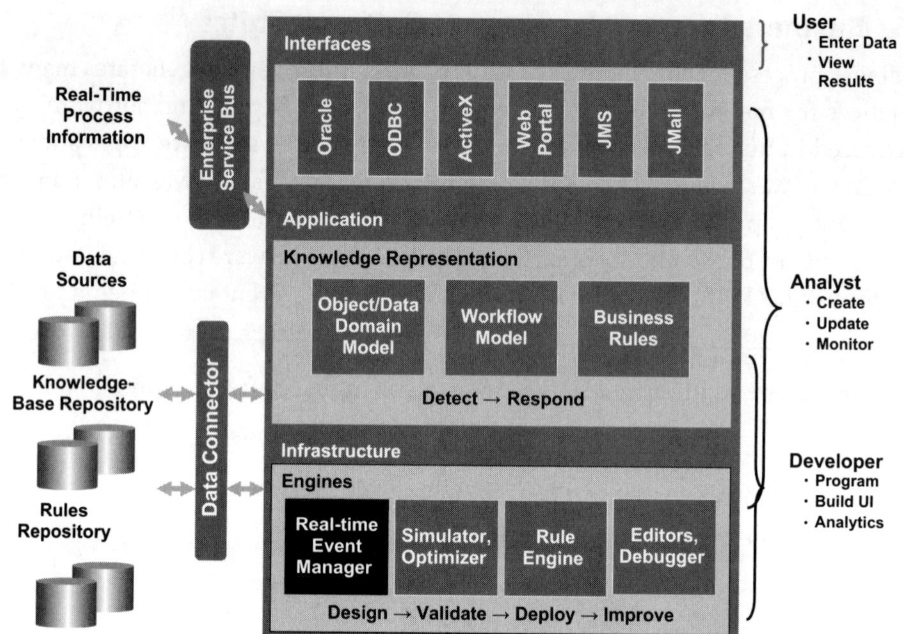

FIGURE 6.6 High-level architecture for BPM with data sources/targets and user roles.

data sources, and targets. Common interfaces are shown in Figure 6.6, and there are many more depending on the vendor's technical approach, applications that may already exist (legacy applications), and desired levels of integration (e.g., with external applications, between internal applications, etc.).

Integration has historically been, and continues to be, a bottleneck in application implementation, including BPM applications. However, the adoption of standards have helped to ease integration because they are flexible and simplify the work required to connect software applications. The combination of improved integration, especially when using the Internet, and software-component technologies (described in Section 5.0 of this chapter) has created powerful synergies in BPM software such as "mashups," a term borrowed from the music industry and defined as the ability to quickly create new applications from the combination of existing components (www.serena.com). For example, one mashup Web site (www.programmableweb.com) allows nonprogrammers to create a Web-based travel log or trip record that uses Google Maps (Google, Mountain View, California) software components to display locations that the user visits, then share this application with others. Business process management mashup capabilities allow business users to create their own internal applications to run reports, create notifications, perform audits, and so forth.

Business process management applications at a high level comprise the application software, business rules, and all data required by the applications. Applications are running programs that process data and include a workflow model that specifies how the work is done and rules that represent decision criteria. Applications are distinct from BPM engines that process business rules (discussed in Section 3.3.3 of this chapter). Business process management systems may include hundreds of active workflows, thousands of rules, and large amounts of data. Business process management applications may be entirely machine-centered or automatic, or they may be hybrids that are partially automated but also require human interaction.

Business process management infrastructure comprises the "engines" that run the instructions provided by applications and software modules that are required across the life cycle of a BPM system. An engine is just like a motor in a car. You give it "gas" (i.e., the rules, workflows, and business data) and it makes the "car" (i.e., BPM application) go. The engines (Section 3.3.3 and Figure 6.6) include a workflow engine, business rules engine, simulation engine, editors and debuggers, and specialized frameworks, for example, for real-time support (handling events, processing time-dependent data, etc.) and optimization. Business process management suites created by vendors include hundreds of optional modules in this layer, depending on the vendor and their particular BPM specialty. Examples include extensions to the features of the rules engine to support "capture" of business rules, optimization modules for inventory control, and built-in customer relationship management business processes.

3.3.2 *Process Editors and Rule Editors*

An important component of a BPM suite and an essential mechanism for knowledge representation is the process and rule designer. The user interface for a rule editor is shown in Figure 6.7, which depicts two editors containing two rules that are used to accept a credit application. The condition part of both rules is the same, but the action part is different; one concludes that the credit application should be accepted, whereas the other posts a message. This is a structured language editor that constrains the way rules are presented and the options available to the user when modifying the rule. Rule editors also allow for rule analysis, such as searching for rule conflicts (same conditions, different actions), collisions (same actions but different conditions), and redundancies (same conditions and actions).

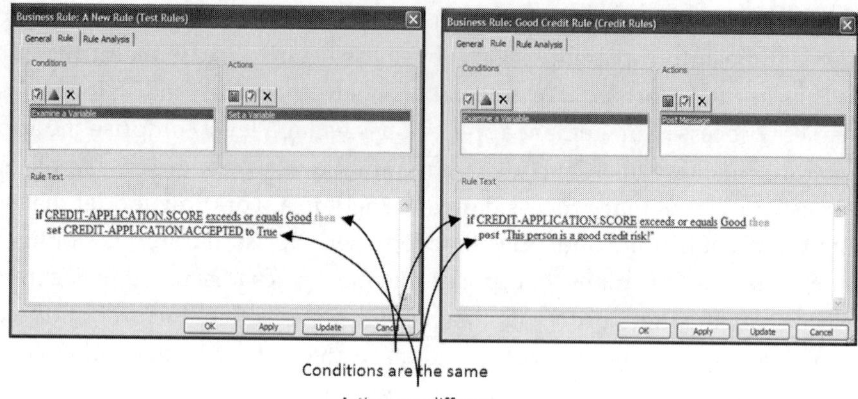

FIGURE 6.7 Two rule editors with the same condition, but different actions.

The condition parts of rules take the form "attribute-operation-value"; for example, the credit application score (i.e., attribute) is (i.e., operation) high (i.e., value). The action parts of rules take a similar action-attribute-operation-value form; for example, conclude that (i.e., action) the credit application accepted (i.e., attribute) is (i.e., operation) true (i.e., value). These characteristics allow rules to be organized in "decision tables" such as that shown in Figure 6.8. Any number of rules, either individual or in groups as in the decision table, compose "rule sets" designed to represent decision knowledge.

Individual rules or rule sets can be invoked at decision points in a process like those processes shown in Figures 6.1, 6.3, 6.4, or 6.5. The association between process activities and rules is created within a process designer or editor like that shown in Figure 6.9.

3.3.3 *Execution Engines*

A BPM suite contains many different engines for processing the instructions represented by rules, workflow, or computer code in the application layer of Figure 6.6. The set of engines in a given vendor's BPM suite varies because each vendor has specialization; however, all good BPM suites will have the following four essential components:

- User interaction, including monitoring and collaboration;
- Statistical/mathematical tools, including statistical analysis, simulation, business intelligence, optimization, and statistical modeling tools;
- Execution, including for rules, processes, and other code; and
- Integration, including that based on standards.

One rule per column

	Label->	PPO Network
	Category->	Auto-pricing
	Priority->	6
Attribute	*Operation*	For any Sunshine-Claim Claim
Auto-pricing	is	yes
Network-type	is	unknown
PPO-indicator	is	yes

Action	Attribute	Operation	
Conclude	Network-type	is	PPO
Conclude	Route-to-CSR	is	false
Start			price (the client of Claim, the symbol PPO)

FIGURE 6.8 A decision table containing a single rule (far right column).

User-interface engines provide a means for displaying information and enabling user interaction. These interfaces may include facilities for sharing information with others, three-dimensional viewing and other visualization (e.g., a geographic information systems [GIS] user interface), a Web-based "dashboard" for monitoring performance metrics in real time, or a collaboration environment for conducting meetings that connect people who are geographically dispersed.

Statistical and mathematical engines are a broad class of tools that share the common feature of being an algorithmic means of generating answers to important business questions such as

- (Simulation) What will happen to revenue in the next quarter if sales go up 25%?
- (Optimization) What's the best reorder frequency given fluctuating, uncertain demand?

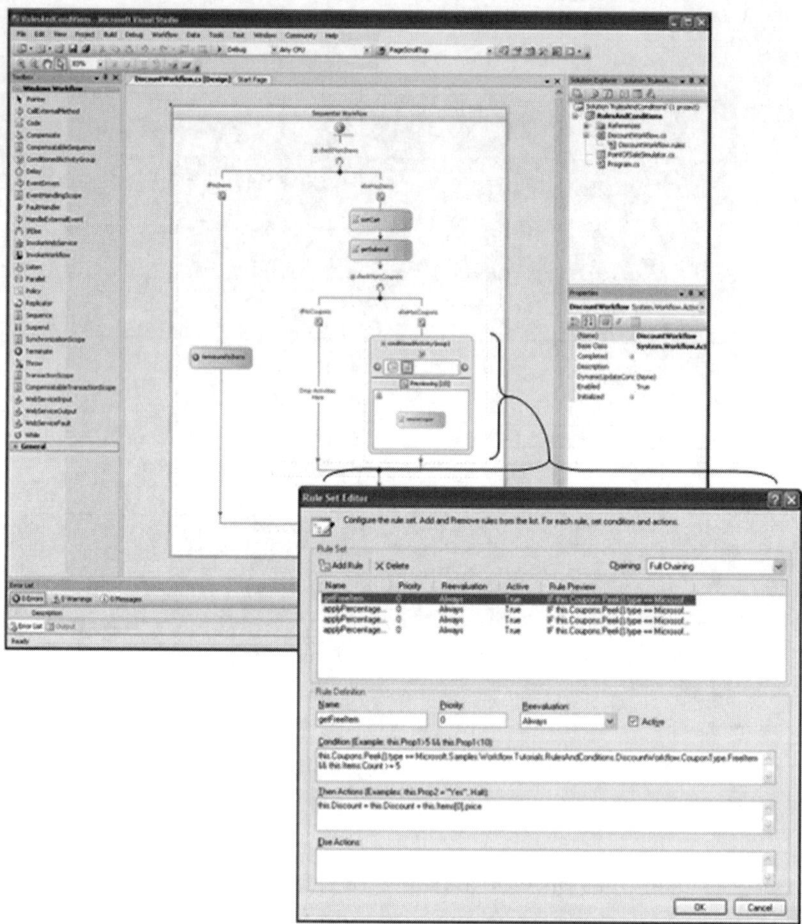

FIGURE 6.9 A process editor (top) and a rule set editor for a process activity.

- (Business intelligence, statistical) Is the performance of third-shift operators significantly different from those on first shift?
- (Business intelligence) What's the root cause of our problem meeting quality targets?

Execution engines do the processing of instructions and provide support for IT requirements such as scalability, flexibility, reliability, and performance. Business process management suites are based on general purpose programming platforms

(Java [Oracle, Redwood Shores, California]; and Microsoft [Microsoft Corporation; Redmond, Washington] languages) from which they derive these characteristics.

3.3.4 *Integration and Architecture*

Efficient integration of applications requires lots of coordination, which explains why this aspect of BPM, like all other software, can be a bottleneck. Diversity of application designs and vendor-specific (proprietary) data management systems, with differing expectations for how data should be viewed and exchanged, makes coordination across systems difficult and costly to implement and maintain. Integration is dependent on the type of software systems the integration must support (see Section 4.0 of this chapter).

3.3.5 *Knowledge Repository*

Business process management theory provides a well-organized, time-tested framework for structuring business information and knowledge, and does so in such a way that it can be easily reused and rapidly automated. One can quickly see that if an organization's processes and rules are modeled, automated, and easily accessed, that a repository of processes and rules is an extremely valuable asset. In some companies, these assets represent all of the value in the company. Because of this fact, BPM suite vendors have emphasized developing and maintaining knowledge repositories.

Repositories are databases that hold information contained in the knowledge representation layer of Figure 6.6. Specifically, this includes the enterprise data model, process workflows, and decision rules. Content in the repository is developed using the tools described in Section 3.3.2. Knowledge in the repository is managed on an ongoing basis by configuration management tools that enable sharing, versioning, auditing, and tracking of knowledge on the basis of workflow, rule or rule set, author, maturity level, division of the company, and so forth.

The user interface for a rules repository is shown in Figure 6.10. This interface shows the rule-attribute information in the top-left panes; one rule being edited in the center showing the rule syntax options, notes, and authors; and the rule display at right. Repository interfaces may have different functions for different user roles (e.g., administrator, developer, analyst, and operator); however, the designs stress accessibility at all role levels to ease development and encourage use by nonprogrammers.

3.3.6 *Monitoring*

Monitoring of BPM systems includes user interfaces for displaying information from running business processes. A distinction is made between the displays and

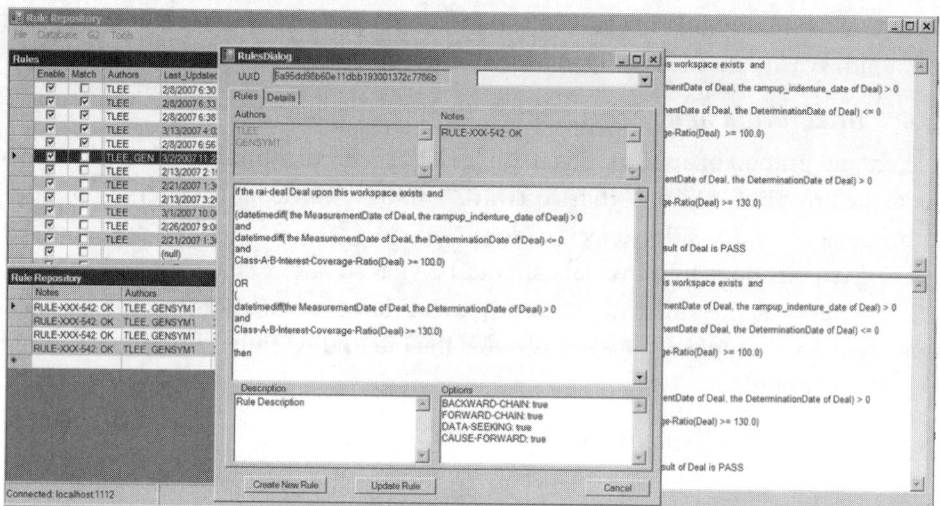

FIGURE 6.10 The user interface for a rules repository.

the underlying analytical engines that may support the display. Data displays may be updated in various ways, including on a periodic basis (e.g., every hour, every day, etc.), when requested by a user, or when an event occurs (e.g., when a new order arrives, a transaction occurs, etc.). *Business activity monitoring* is a term that was coined to describe monitoring applications that provide real-time updates of business metrics from running processes. Business activity monitoring takes advantage of the analytical tools described in Section 3.3.3, including business intelligence engines, optimization, simulation, tracking or auditing, and more.

3.4 Business Process Management Frameworks

Business process management frameworks combine software architecture with a methodology aligned to that architecture. Frameworks are created by software vendors, consultants, and independent standards organizations to promote effective practices. This section provides an overview of simple software architectures and methods that have a track record of success in both small and large organizations. Software architecture and methods in this context support and implement the enterprise-level IT architecture, which is the subject of Section 4.0 in this chapter. This section describes BPM frameworks in the context of an enterprise BPM system. An enterprise BPM system is used as an example, but a smaller division or group within an enterprise could use the same approach. Business process management frameworks describe BPM systems in the same way that blueprints describe the design of a building. Business process management methods describe how to create the blueprint and, like construction practices, specify step-by-step procedures for getting from the blueprint to actual BPM implementation. Business process management frameworks also support continuous improvement functions so that business processes can become more effective and efficient.

3.4.1 *Value-Stream Analysis*

Developing an enterprise business process model begins with value-stream analysis. Historically, value-stream analysis was used as a way to identify functional units in an enterprise. The value perspective is taken (rather than, for example, a more limited perspective such as cost) because creating value is central to the mission of all enterprises. Value encompasses all other important factors such as cost, safety, security, and so forth. Each unit of the company must be justified in some way in terms of generating value, either directly or indirectly.

Value chain analysis leads to identification of functional units common in an enterprise. Different industries emphasize different functional units depending, more or less, on how they generate value (e.g., who the customers are, their particular demands, and the enterprise infrastructure needed to meet customer demand). For example, telecommunications companies focus on provisioning new mobile phones and billing for services. In addition, manufacturing companies focus on production efficiency to meet quality and quantity targets. Both of these industries have developed reference models that identify important functional groups of business processes (e.g., provisioning and billing for telecommunications and production for manufacturing).

Reference models are hierarchical models, with aggregated functions composing other functions at greater levels of details (see Figure 6.3). Each function can be represented as a business process and the decomposition of a function can be shown as interconnected sequential or parallel business processes. The real utility of a reference model lies in the model's definition of best practices and standard metrics for business processes at all levels of the hierarchical business model. Best practices lists enable business process designers to choose effective process implementations. Because they are logically linked and tied directly to specific business processes, metrics enable determination of the cause(s) of performance problem(s). Process definitions, their interactions, best practices, and metrics are all the ingredients a methodologist needs to build a blueprint, design business processes, and implement continuous improvement.

Business process management practitioners use reference models to identify the business processes in their organization and to create business process models of their enterprises, or an *enterprise business model* (EBM). If a convenient reference model does not exist then relevant parts can be borrowed from models used by related industries. For example, water and wastewater utilities are, in many ways, like telecommunications companies because they focus considerable energy on customer service and billing. Water and wastewater utilities are also like manufacturers because they provide both products and services (collection, treatment, and transport). Reference models from these industries provide valuable ideas that, together with in-house expertise, can be used to construct an EBM blueprint tailored to the unique way in which a water or wastewater utility creates value for their community.

3.4.2 *Business Process Management and Continuous Improvement*

There are three common approaches to continuous improvement of BPM systems, each growing from separate disciplines (Business Process Trends Home Page, www.bptrends.com). There are the business management disciplines that are focused on process reengineering within the context of enterprise value generation, much as described in Section 3.4.1. A second discipline is focused on quality control and work process simplification, historically for manufacturing processes but now also for any type of work process. The third discipline was developed by computer programmers and software analysts, who emphasize the importance of IT as an enabler. Each of these disciplines is supported by different groups of professionals with supporting methodologies and tools. All are relevant to BPM continuous improvement and there is a natural ongoing evolution that is resulting in consolidation of these disciplines.

The business management discipline is exemplified by the work of Hammer (1990). Emphasis is placed on corporate strategy, particularly in creative ways, by taking a process-centric, holistic view that considers all levels of the organization. Reengineering proponents eschew technology as the resolution, insisting that "technical problems are the easy problems." Instead, reorganization of people and processes is central, everyone is involved, and change management processes are instituted to move the enterprise to a new way of creating value. Although the methods have evolved to be less radical, they are still process-centric and emphasize alignment of people, processes, and technology. They balance total redesign and reuse of existing processes, incorporating methods from the other two frameworks.

The quality control/work simplification discipline evolved from manufacturing industries. This tradition includes Six Sigma and Lean Production (Toyota, Tokyo, Japan) methods, both of which make extensive use of statistical process or quality control data to identify areas for improvement and cost savings. Six Sigma's stepwise Define, Measure, Analyze, Improve, and Control process provides concrete steps that organizations can follow to continuously improve. Similarly, Lean Production techniques focus on eliminating waste in the expenditure of resources used to create value. Lean Production looks for seven or eight specific types of waste and provides best practices for identifying and eliminating waste.

The IT-centric approach is largely an enabler of the other two disciplines, but has been prominent because BPM often has a goal of automation of business processes. The BPM suite (Figure 6.6) represents one important artifact from this discipline, that is, consolidating workflow, rules, enterprise integration, the Internet and many other technologies for the purpose of implementing BPM. Methods that exist are standardized only in the sense that practitioners select a specific BPM suite, including extensions or different combinations of technology shown in Figure 6.6, then construct methods for implementation of enterprise business models around that BPM suite. The benefits of this can be significant; for example, the value of applying rules technology as described in Section 3.2.3 empowers business managers by enabling them to control the rules that result in value creation within their enterprise.

3.5 Trends in Business Process Management

Business process management continues to evolve along at least four dimensions: in the development of best practices or successful patterns of BPM application, in the degree to which BPM reaches into organizations, in the degree of dynamism in BPM, and in the advancement of software technology to support BPM. Process engineers in water and wastewater utilities will recognize the similarity between these trends and those observed in control of the engineered unit processes in their plants. This is not a coincidence as BPM is all about managing processes.

Identification of many business processes using the techniques and tools described in this chapter have resulted in a proliferation of ideas on standard business processes, the form of rules that drive decisions in a business process, and integration between processes. Practitioners of BPM have begun to see many patterns of application that are more effective when compared to alternatives and variants that meet efficiency, cost, security, and other requirements. Outsourcing of business processes to a third party is a strategy for creating efficiency when those business processes have become standard (e.g., have become commodities like paper clips) and can be purchased inexpensively. Standard patterns exist in commercial software products with out-of-the-box business processes that can be used. For example, common small business accounting software contains workflows for order processing, banking transactions, and so forth. These workflows are well-established from generally accepted accounting procedures. Any small business can have access to these processes for the cost of the software. Patterns also show up in industry reference models, as described in Section 3.4.1. of this chapter.

The reach of BPM refers to the degree to which this theory has penetrated organizations. BPM started in areas where the needs were greatest. For example, customer-facing business processes often get attention because most organizations exist to serve customers. SalesForce.com and other customer-relationship management companies have grown significantly because they offer business processes that are standardized and optimized. In the manufacturing industry, quality control processes drive both quality and cost to create better products that increase company revenue. In the telecommunications industry, provisioning and quality of service are highly refined as anyone with a mobile phone quickly appreciates. Business process management is growing into all levels and divisions of organizations, bringing the potential for improvement.

A major trend in BPM is the greater appreciation for time dynamics in business processes. Many business processes are static. A static business process cannot adjust to a change resulting from events that occur in time. The workflow is fixed, data that change with time are averaged to remove variance, and activities are processed sequentially in one shot without provisions for stopping, rolling back, or feeding back information. In contrast, dynamic business processes are like the controllers managing dissolved oxygen concentrations or chemical additions to plant processes. Real-time BPM incorporates events that occur in time and mechanisms for responding to events. Dynamic BPM increases the flexibility and responsiveness of an organization. Understanding time effects enables the organization to adapt when the environment signals a change. The overall result is a much more agile business, that is, one that is aware of changes in the business environment and has the ability to modify its processes to adjust to change.

Software and other technology for BPM (see Section 3.3) changes rapidly. These industries have been, and will continue to be, highly dynamic. However, the organization of software components in a BPM suite, as shown in Figure 6.6, will be maintained. Areas where improvements have been observed were discussed in Section 3.3.1.

4.0 INFORMATION TECHNOLOGY AND ENTERPRISE ARCHITECTURE

4.1 Introduction

Business process management and enterprise architecture share some common objectives and methods. Both define systems in terms of business processes and both enable IT's alignment with business goals. Whereas BPM focuses on business systems and analysis, enterprise architecture focuses on exploring and describing the IT implementation of such an analysis. Enterprise architecture benefits from BPM in that the necessary analysis and conversation about business goals, priorities, strategies, and constraints has created a foundation from which the architect can begin IT system tasks. In addition, what they have in common is a growing body of knowledge, practices, tools, and frameworks that promote effective decision making and understanding across the organization.

The term, *architecture*, occurs throughout this manual of practice and, depending on the context, involves different activities, responsibilities, and roles. Architecture, in all contexts, foresees the goals of a system and models and constrains a system to meet present and future needs of the system's users. The nature of the system determines the level of architecture. For example, a small system, specific to a discrete business task, will have a less abstract architecture, whereas a large system, like the entire enterprise itself, will be defined and described by a higher-level, more abstract architecture. Information technology architecture and enterprise architecture are often synonymous; however, IT architecture scales down into implementation specifics, whereas enterprise architecture scales up toward long-range planning.

Other commonly used terms in the modern business lexicon (e.g., *program*, *project*, *system*, *design*, *process*, etc.) occur at both the lowest and highest levels of the organization. Like these terms, *architecture* is also commonly used. It covers tasks, practices, designs, intentions, plans, or structures that apply to many interests, both large and small. These multiple-use contexts imply imprecision, yet flexible terms often support abstract concepts. In computer science, *function overloading* describes the practice of generalizing similar logical operations under one function name. As such, various functions using the same name can cover multiple conditions. In this way, we can say that *architecture* (and *project*, *system*, *process*, etc.) is an overloaded term. These terms generally convey consistent logical meaning within a given context. In addition, there are different types of architectures just like there are different types of systems. To help sort this out, an architecture hierarchy is described in the following section.

4.2 Architecture Hierarchy

Enterprise architecture defines the direction and priorities for all business information systems. This type of architecture supports the organization's portfolio of systems and prescribes which programs get funded and how. Next are the architecture frameworks (discussed in Section 4.5.3), led by IT to frame the current and future information systems environment. This level is commonly known as *IT systems architecture*, although it overlaps the domain of enterprise architecture. At a greater level of detail and specificity and a narrower focus is application (software) architecture, in which specific systems are designed, constructed, and customized. This architectural context applies during specific project planning and execution, often led by a technical leader called the (software or system) architect. Next, there are function-specific architectures like network architecture (also known as network topology), security architecture (policies, hardware/software devices), and storage architecture (devices for data storage, backup, connectivity, etc.). Architectures exist almost everywhere.

Beyond the organization itself, IT architectures can convey community, policy, technical trends, and design principles. This includes open architecture (as mentioned in Chapter 7), which embodies public standards; service-oriented architecture (SOA) (Chapter 6) for system components integration; model-driven architecture (MDA) (Chapter 6) for advanced programming methodology; and others.

This section emphasizes architecture at the strategic level, incorporating and weighing in on long-term effects and decision making. The concepts of architecture as a design specification (as in construction) and architecture as a planning process, aligned with business strategy, are revisited.

4.3 Information Technology Architecture

According to The Open Group (San Francisco, California), an industry standards organization, "IT architecture defines the components or building blocks that make up the overall information system. It provides a plan from which products can be procured, and systems developed that will work together to implement the overall system. It thus enables you to manage your IT investment in a way that meets the needs of your business" (The Open Group, 2004).

As in any complex structure, like a modern home or office building, IT architecture defines a framework in which the system (the structure) exists, both *before* it exists and at various times during construction and maintenance, as follows:

- *Boundaries*—walls, property, floors, conduits, stairwells, and so forth;

- *Constraints*—foundation, materials, height, and so forth;
- *Subsystem interactions*—ducts, electrical to heating/ventilating, security, monitoring, and so forth; and
- *Overall characteristics*—style, décor, usability, convenience, efficiency, extensibility, and so forth.

Further, like a building's architecture, which meets the occupants' needs and goals, it is the purpose of IT architecture to envision the complete system at multiple levels of detail, providing sufficient specificity to enable others to implement it over time (i.e., a long-term framework for planning, procuring, developing, installing, and maintaining an information system that meets the goals of the organization). Although IT architectures do not solely prescribe physical properties like a building's architecture does, the analogy works from a conceptual point of view and represents a convenient model from the physical world.

4.3.1 *Information Technology and Enterprise Architecture as a System for Systems (Meta System)*

Good architecture represents the organization's vision; it is based on a strategic, top-down standpoint for IT, not a bottom-up, technical view. Consider IT architecture as a high-level blueprint for describing and managing enterprise IT infrastructure and rules in support of business objectives and policies. At the high level, it suppresses implementation details and vendor-specific solutions and expresses capabilities and policies.

The Institute for Enterprise Architecture Development (Amersfoort, The Netherlands) sums up the key guiding principle of the IT and enterprise architecture discipline as follows: "No strategic vision, no [enterprise architecture]." In other words, today's enterprise architecture is about delivering tomorrow's business systems. An important aspect of this assertion is that enterprise architecture is a holistic discipline that unites business and technology elements based on a strategic enterprise vision (Temnenco, 2007). In this sense, enterprise architecture establishes a systematic way of looking at all the systems that comprise the organization's information services.

4.3.2 *Measure Twice, Cut Once*

Beyond a set of practices or techniques, enterprise architecture provides a strategic planning context for the evolution of IT systems in response to the goals of the organization. Enterprise architecture represents "what is" and anticipates the "to be." It is an abstraction of the IT infrastructure for control and planning and an integration of IT and business; it supports capital planning and IT investments. The forward-looking, planning aspect of enterprise architecture reduces waste by reducing the number of ad hoc decisions that can have costly consequences. "Measuring twice" refers to considering multiple courses and alternatives before committing to a particular IT decision and it also means evaluating potential solutions thoroughly.

Enterprise architecture incorporates operational risk management and how the business strategy is transformed into an IT strategy. The architecture is both a tool for decision making and a reflection of those decisions that will guide and anticipate specific IT implementations.

In general, IT decisions often have long-term consequences for future business and operations. When they are made without the guidance of an overall architecture, the organization incurs additional long-term risks from potential technical limits that can affect many dependent systems. A good architecture plan is a measure of an organization's capacity for meeting its goals and objectives.

4.3.3 *Information Technology and Enterprise Architecture Manages Complexity*

Enterprise IT involves complex systems with multiple technologies supporting many business processes. Each system (e.g., billing, customer service, and work order management) may have its own infrastructure, primary user group, support system, and vendor relationship. Implementations and changes to these systems occur at the project level and are typically managed as "solutions." Projects and solutions have plans, technology choices, and architectures as well. They occur as complex activities undertaken by IT specialists who "speak the language" of IT implementation. In the final analysis, there is no other way to "do" IT.

Ideally, IT solutions are derived from the enterprise architecture, which itself is derived from a business plan and maintained in concert with business objectives and priorities. Managing and prioritizing implementation projects must follow the lead of the enterprise architecture to ensure that the solution generates the expected business benefits. Making sense of detailed implementation plans occurs within the larger context of enterprise architecture. The enterprise architecture context is holistic and its perspective is organizational and strategic, whereas the solution architecture is implementation-specific (Temnenco, 2007).

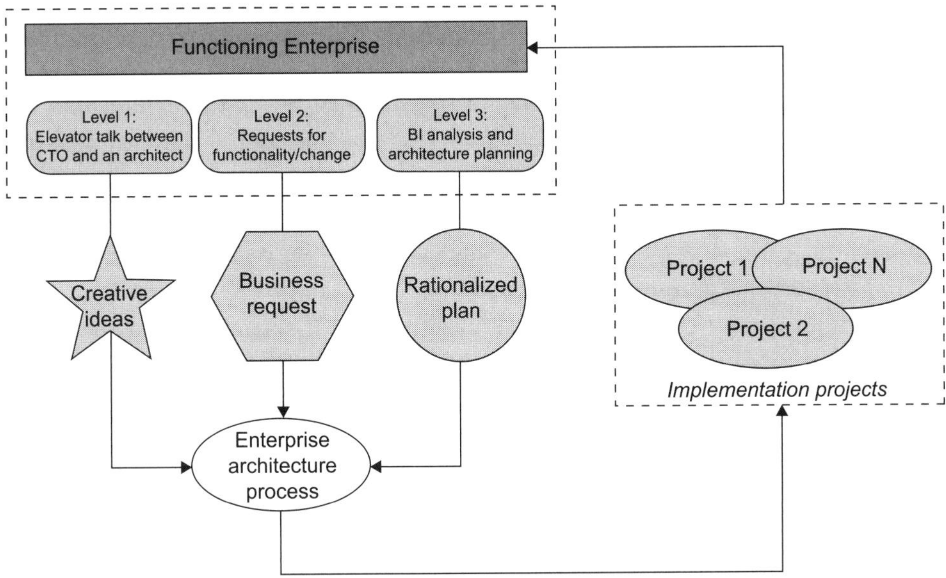

FIGURE 6.11 Projects that derive from the enterprise architecture (Temnenco, 2007).

Figure 6.11 shows the relationship between IT projects and the enterprise architecture; it is the enterprise architecture perspective that sets priorities for project decisions.

4.3.4 Information Technology and Enterprise Architecture Manages Boundaries

An enterprise architecture encompasses multiple, large subsystems. Some subsystems, such as customer service, billing, and a laboratory information systems (LIMS), meet the needs of specific organizational units; other subsystems, such as e-mail, network, phones, and Web sites, serve the organization as a whole. Generally, enterprise architecture recognizes organizational boundaries by a combination of policy and technology. It meets organizational needs across subsystem boundaries by

- Identifying subsystems and their dependencies;
- Documenting interactions, compatibilities/incompatibilities, and access rules across subsystems;
- Reducing and isolating risks across communicating systems; and
- Describing subsystem contexts and defining the overall boundaries.

4.3.5 *Information Technology and Enterprise Architecture Manages Change*

In terms of managing boundaries, enterprise architecture accommodates and constrains organizational change. The forces driving organizational change come from many sources (i.e., political, economic, customer expectations, technology, staff, resources, etc.). Good enterprise architecture enables the organization to achieve the right balance between IT efficiency and business innovation as it responds to change.

Good enterprise architecture also allows individual business units to innovate safely in their pursuit of organizational objectives. At the same time, it assures the needs of the organization for an integrated IT strategy, permitting the closest possible synergy across the enterprise (The Open Group, 2004). It also allows for better return on existing investment and reduces the risk for future investments. Finally, related activities like procurement are simpler because the information is readily available in a coherent plan.

4.3.6 *Information Technology and Enterprise Architecture Manages Opportunities*

Besides enabling effective change management, well-designed enterprise architecture supports a more nimble organization, that is, one that can better assess risks and potential rewards in new opportunities. Awareness of, and planning for, the state of enterprise architecture enhances business knowledge and decision making. Managing opportunities includes the ability to identify cost savings, improve service, reuse existing systems, and respond to unforeseen needs.

4.4 Themes in Water and Wastewater Utilities' Information Technology Architecture

4.4.1 *Technical Themes: From Mainframe to BlackBerry®*

Many utilities' IT systems predate the Internet age and are characterized by combinations of legacy systems operating in tandem with newer systems. Legacy systems represent long-term investments, require specialized skills that are harder to find, and suffer from a lack of capabilities that make them expensive and inflexible. However, they are not easily changed because of many years of infrastructure and business processes that have grown up around them.

The history of computing has gone from the mainframe (the air-conditioned "glass room"), to mini computers, to desktop computers, to networked computers and servers, and, currently, to Web-based (Internet) applications using "thin clients" and powerful "back-end" personal computers that mimic the original mainframe as a central repository of computational power and control. In addition, there are a plethora of post-Web-era mobile devices such as BlackBerrys®, toughbooks, e-books, radio frequency identification, and global positioning systems. Innovations continue to emerge that leverage increases in computing power and generate entirely new modes of functionality in an ever-expanding, always-on computing environment.

Besides the expanding array of interconnected computing devices, massive improvements have carried over into industrial systems, such as remote sensing, fiber optics, automation systems, wireless access, and monitoring and measuring equipment. For example, SCADA architecture was formerly one personal computer running a single human machine interface (HMI) software license for development, data collection, and viewing (client). Redundancy was poor and networking was proprietary per vendor. Remote access to the control system was strictly forbidden as a security risk. Now, however, the following factors are commonplace:

- A push toward integrated architecture, that is, not a single personal computer anymore, but multiple servers such as data collection (SCADA) servers, historical servers (and databases), more sophisticated reporting, and data analysis demands;
- Use of server class machines with increased configuration complexity;
- Support for remote access of SCADA HMI screens;
- Switch from proprietary networks and protocols to almost exclusively transmission control protocol /Internet protocol (TCP/IP) over Ethernet; and
- All the above factors require coordination with the IT infrastructure because of tight integration with networks, firewalls, routers, terminal services, remote desktops, etc.

These advances in capabilities and capacity at lower costs (and higher complexity) drive, and are driven by, the architecture when selecting or upgrading systems. In addition, the layers of older and newer technologies in most utilities create complex infrastructures that reflect the long life of architectural decisions. The "measure twice" building architecture analogy described in Section 4.3 again illustrates how difficult it may be to undo a foundational design.

4.4.2 *Technical Standards*

The emergence of the Internet in the 1990s brought global standards for computerized networks that accelerated change and fostered trends that continue to affect enterprise architecture. These include

- Interoperable, standards-based systems (as opposed to proprietary, vendor-based systems that create vendor "lock in");
- The increased role of standards-making bodies like the World Wide Web Consortium (W3C), Organization for the Advancement of Structured Information Standards (Boston, Massachusetts), the Open GIS Consortium (Wayland, Massachusetts), the Object Management Group (OMG) (Needham, Massachusetts), The Open Group, European Computer Manufacturers Association (Geneva, Switzerland), and so forth;
- Open-source software (e.g., Linux operating system; Apache Web server; and PHP, PERL, and Python Web scripting languages) and new software licensing models;
- Web services that allow disparate computer systems to interact via XML-based messaging protocols;
- Software as a service (SAAS), which delivers applications to customers via the Internet, eliminating the installation and maintenance cycle entirely (like Salesforce.com, Google applications, and a growing number of conventional software vendors like Microsoft, Oracle Corporation [Redwood Shores, California], and SAP [Newtown Square, Pennsylvania]); and
- "Cloud computing" and SOA, which, like Web services and SAAS, build upon Internet connectivity to deliver services and systems from outside the customer's premises, reducing the need for capital-intensive, internal systems.

4.4.3 *Global Culture of Technology*

Pervasive use of the Internet has also raised end-users' expectations for real-time access to information at any time and created new communities of collaboration. In addition, entirely new human-to-human interactions are built on the Internet infrastructure (e.g., social networking, instant messaging, and texting), the convergence of phone and Internet, location-based services, and mobile computing. Tracking these changes and incorporating the best new enabling technologies into an existing organization's computing ecosystem naturally challenge many C-level managers and their staffs. No legacy architecture can keep pace in the Internet age.

The plethora of computing devices, interactive networks, and global communications links generates other factors affecting architectural decisions. These are globalization of the workforce, outsourcing, and security. As an example, instead of replacing an inefficient billing system, consider outsourcing as an option (via SAAS, an outsourced remote call center, or bill processing operation).

4.4.4 *Security*

Security concerns today are not as simple as they used to be. Indeed, security today covers everything from September 11-induced first-responder requirements, to malicious viruses and rogue programs, to sabotage and massive identity thefts. Enterprise architecture that does not factor in a well-crafted security policy will not protect the organization's interests for long because security must be built in to planning and decision making.

4.4.5 *"Green" Values*

Sustainability and environmental stewardship are new drivers in the design of information systems. It is no longer sufficient to be cost-conscious; according to the U.S. Department of Defense (U.S. DOD) (2007), it is important to be "aware of cost effective acquisitions that achieve compliance with policies, reduces resource consumption and solid and hazardous waste generation." The growing green consensus is occurring across multiple constituencies: consumers, investors, shareholders, and regulatory agencies. "Practicing green procurement demonstrates an organization's commitment to considering and minimizing the environmental consequences of its activities. It thus makes both environmental and economic sense" (BSDGlobal, 2009).

Some "green" topics that affect architecture, design, and operations are green procurement and total life-cycle management of digital assets, energy efficiency (e.g., smart devices that monitor use and adjust to peak/off peak demands and smart grids), alternative energy sources such as wind and solar, and byproduct disposal and reuse (e.g., biosolids management programs).

These are the many themes underlying today's IT and enterprise architecture challenges: continual technological advances, cost-conscious rate payers, demands for service in Internet time (24 × 7), globalization of resources, secure and safe computing, doing more with less, and protecting the planet, all while anticipating new demands for service and efficiency.

4.5 Methodologies and Techniques for Developing Enterprise Architecture

4.5.1 *The Enterprise Architect*

This role of enterprise architect typically exists within the IT business unit; it provides technical thought leadership and primary ownership of the technical vision and understanding of the enterprise system. Because architecture must embody the goals of the enterprise, the enterprise architect uses strategic relationships with peers and leadership in both IT and business management. The enterprise architect is responsible for documenting both the "as-built" state and defining the technical roadmap that supports the short- and long-term strategic objectives of the business.

4.5.2 *Approaches to Developing an Enterprise Architecture*

As the "systems must be driven by the needs of the enterprise, not by technology... this requires that the business be modeled before attempting to implement its systems in software" (Marshall, 1999). The process of defining enterprise architecture includes creating models from different perspectives that show existing system(s) in the current organization (the as-is views) and other models that show how the architecture extends to meet future needs. The enterprise architect builds the information system models upon the foundation of business strategy.

4.5.2.1 The Roadmap: High-Level Technical Vision and Functional Requirements

Beginning with the strategic vision establishes the long-term framework for the overall enterprise IT mission. From that foundation, the architect must work with stakeholders to prioritize high-level goals, weeding out or deferring pet projects and similar nonaligned efforts. The architect defines the as-is architecture in figures and descriptions and compares the as-is to long-term goals to identify the gaps in current capabilities and explore which new systems are needed. Within this process, a project portfolio takes shape that begins to itemize the high-level requirements and features of the new architecture. The to-be architecture not only includes requirements and features, but identifies current and future constraints, which are limits to technology, schedule, funding, staffing, and other resources.

4.5.2.2 Define the Technical Vision with Documents and Models

The output of the preceding activities, undertaken over several iterations with key stakeholders, is a high-level technical vision supported by key attributes of the new enterprise IT model and a long-term plan of prioritized projects that have at least a target range for scope and budget in the form of business documents and diagrams, which may include the following:

- Technology diagrams depicting major hardware and software components (layers) and how they interact;
- User interface and workflow diagrams that illustrate the major business processes as a series of user interactions, use cases, decisions trees, and system boundaries; and
- Domain models (also known as *conceptual models*) drawn from business processes (the "problem domain") and their major subsystems, data entities, and relationships.

There is no single method for generating these artifacts; in the initial stages, in particular, a white board is often the best tool for "getting it down in writing," as shown in Figure 6.12.

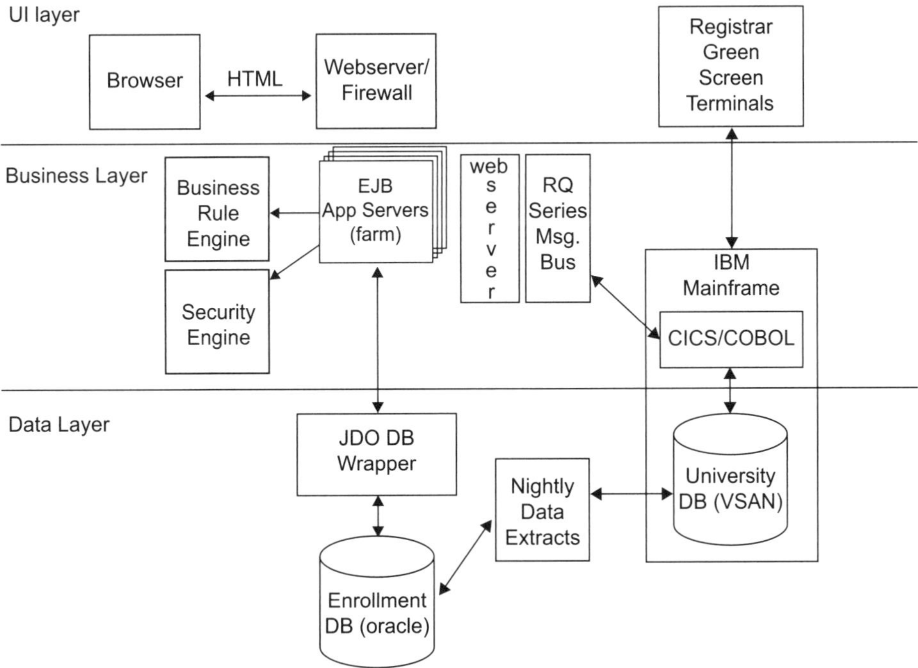

FIGURE 6.12 Technology "stack" draft diagram (Ambler, 2003).

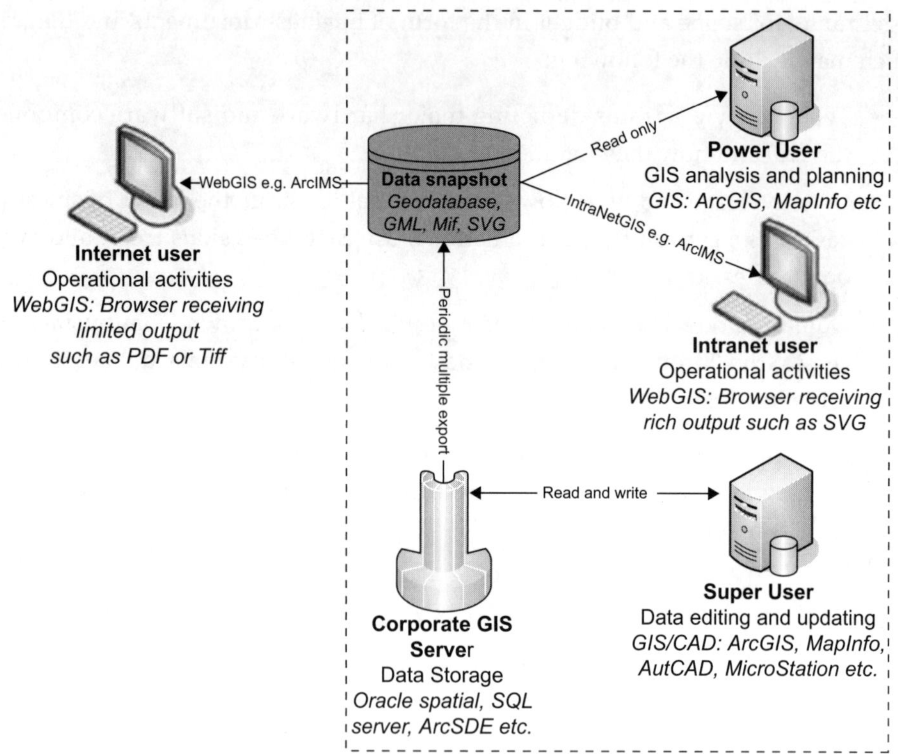

Figure 6.13 Conceptual model of GIS subsystem (Beck et al., 2007).

In other cases, more formal diagrams may use a modeling language (e.g., Unified Modeling Language [UML] or Systems Modeling Language), as described in the Section 4.5.4, or a conceptual diagram, as shown in Figure 6.13.

4.5.3 Methodologies and Frameworks

Because the enterprises or systems described tend to be large and complex, the associated models also tend to be large and complex. To manage this scale and complexity, an architectural framework defines different, but complementary, views of the enterprise or system model corresponding to different stakeholders or areas of interest (Artisan Software Tools, 2008). There are industry-specific architectural frameworks for government and military applications (e.g., U.S. Department of Defense Architecture Framework [DoDAF] and British Ministry of Defence Architectural Framework [MODAF)]). Models for utility infrastructure exist for certain operational subsystems, particularly spatial systems, such as the Spatial Data Standards for Facilities, Infrastructure, and Environment (SDSFIE) (visit the SDSFIE Web site at www.sdsfie.org), and there is research into standardizing operational data models (see Beck et al., 2007); however, a standard framework for utility modeling has not yet been standardized.

There are generalized frameworks for defining enterprise architecture that can be applied to any organization's information management environment. Most define not only a framework of terminology and documentation, but of methodology. These include the following:

- *Zachman Framework*—(discussed in Chapters 4 and 5) is a structural (static) framework that is most effective when used as a model for analysis and classification of artifacts and meta-analysis of methodologies and frameworks (Temnenco, 2007).
- *The Open Group Architecture Framework*—This is a detailed method and set of supporting resources for developing an enterprise architecture. The original version in 1995 was based on a framework developed by the U.S. DOD (The Open Group, 2009).
- *International Electrical, Electronic Engineering (IEEE) Standard 1471–2000*—A recommended practice developed by the IEEE (Piscataway, New Jersey) Computer Society's Software Engineering Standards Committee that establishes a conceptual framework and vocabulary for talking about architectural issues of systems to promote sound architectural practices (see Figure 6.14) (Hilliard, 2000).
- *Software Architecture Analysis Methodology (SAAM) and the 4+1 View Model*—Developed mainly by Philippe Kruchten at Rational Software (IBM Corporation; Armonk, New York), this modeling approach describes software using four concurrent views, each addressing a specific set of concerns. Architects capture their design decisions in these views to illustrate and validate them. The views are the logical view, showing data and/or object oriented analysis; process view, including executable tasks, performance, and communications; physical view, which maps software onto hardware and reflects its distributed aspects; and development view, which describes the software's organization during development (Kruchten, 1995).

In a review of enterprise architecture frameworks, Forrester Research, an authoritative IT research firm, observed that "...it can be very difficult to determine which framework is right for an organization. Forrester's recommendation is to keep it

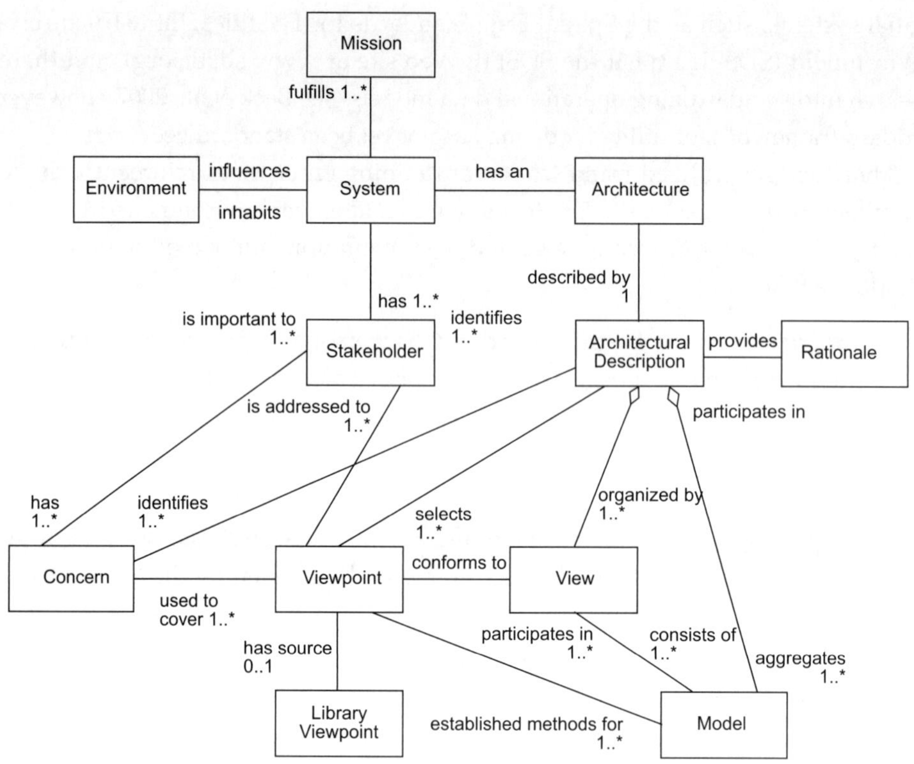

FIGURE 6.14 IEEE 1471 Conceptual Framework for Enterprise Architecture (Hilliard, 2000). © (2000) IEEE.

simple, borrow from the best, and customize for your own organization" (Leganza and Brown, 2004).

4.5.4 *Notational Tools and Documents*

Systemized notation was introduced to model and communicate complex systems for software engineering based on terms and standards from electrical engineering. In more recent times, the need was recognized to develop languages and symbologies that could be used more easily by business analysts. Other standards arose for domain-specific languages (as previously mentioned, the SDSFIE is such a system).

4.5.4.1 *Architecture Notations*

Of the current notations, OMG's UML is probably the most widely adopted and extended as it can fit in so many levels of abstraction or detail. Unified modeling language derives from the principles of object-oriented analysis that describes business systems in terms of entities (objects), operations, and the relationships among them. Without too much training, system architects, business analysts, and interested stakeholders can model systems in any level of detail using very few symbolic types, as shown in Figure 6.15.

An aspect of UML that is useful at higher levels is its support for building customized notation from basic symbols (called *stereotyping*). Figure 6.16 illustrates this using an organization process diagram and enterprise stereotypes (Marshall, 1999).

As mentioned in Section 3.0, another notation for analysts and architects is business process modeling notation (BPMN), "a graphical representation for specifying business processes in a workflow. Business process modeling notation is maintained by OMG. The objective of BPMN is to support business process management for both technical users and business users by providing a notation that is intuitive to business users yet able to represent complex process semantics" (Wikipedia, 2009b) (see Figure 6.17).

Finally, there are self-defined notations that can be developed to suit an organization's communication style showing processes, connections, activities, data storage, hardware, and software that can be used in general situations. As no widely adopted, standard notation has emerged, it may be sufficient for a program manager

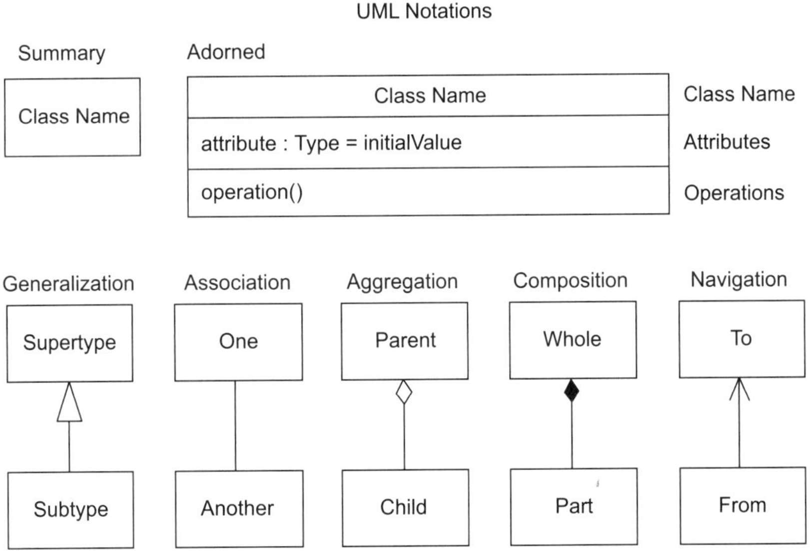

FIGURE 6.15 Unified modeling language notation types (Marshall, 1999. [Marshall, SOFTWARE COMPONENTS FOR THE ENTERPRISE: BUILDING BUSINESS OBJECTS WITH UML, JAVA, AND XML, © 1999 Addison Wesley Longman, Inc. Reproduced by permission of Pearson Education, Inc.]).

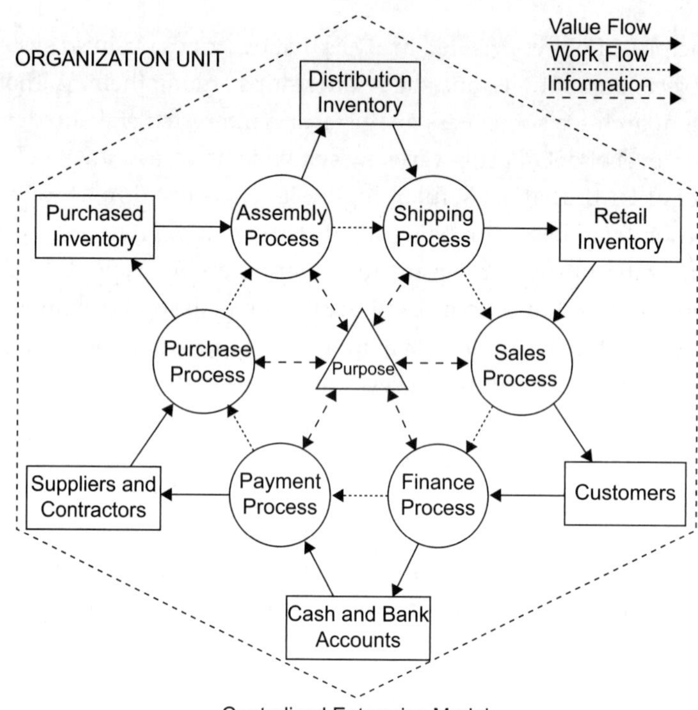

FIGURE 6.16 Enterprise modeling with UML (Marshall, 1999. [Marshall, SOFTWARE COMPONENTS FOR THE ENTERPRISE: BUILDING BUSINESS OBJECTS WITH UML, JAVA, AND XML, © 1999 Addison Wesley Longman, Inc. Reproduced by permission of Pearson Education, Inc.]).

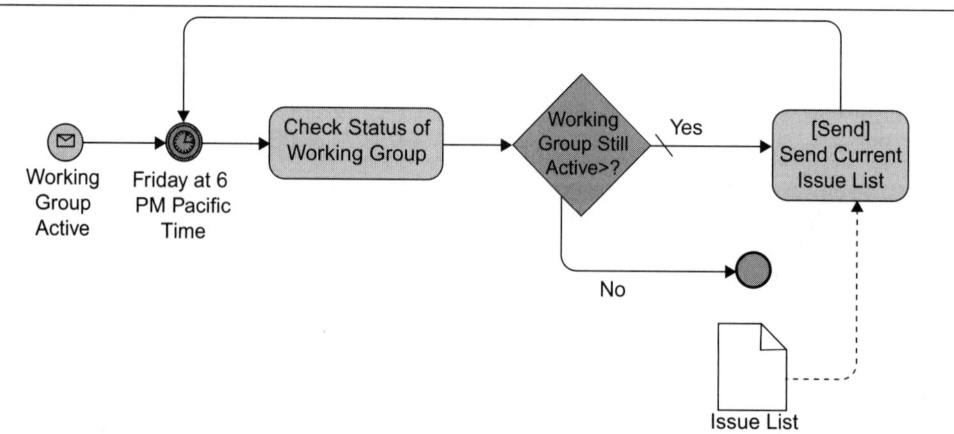

FIGURE 6.17 Example BPMN diagram (Wikipedia, 2009b).

to "do it yourself." For some, a typical representation distinguishes between components that are active (i.e., transform data) and passive (i.e., store data). Groupings of components may be depicted in subsystems or layers in addition to the allocation of software to hardware. This type of notation is typical of ad hoc "white boarding."

4.5.4.2 Tools

Besides using a white board and generic drawing tools, specific tools for modeling enterprise processes and domains include Rational Rose (IBM), Visio (Microsoft), Enterprise Architect (Sparx Systems), and Telelogic System Architect (IBM). (Visit these respective vendors' Web sites for more information.) All provide the means to generate standard figures, diagrams, documentation, and linkages to different levels of detail and abstraction that can make maintaining the documents semi-automated.

No tool replaces the enterprise architect's own abilities to analyze the organization's IT mission and generate a suitable model at a suitable level of detail. It is the enterprise architect that translates the organization's IT infrastructure, constraints, and plans into a useful management document that facilitates communication, prioritization, and procurement at all levels of the organization.

4.6 How to Evaluate Appropriate Information Technology and Enterprise Architecture

According to the SEI (2009a), enterprise architecture can be assessed to meet a variety of objectives that meet the concerns of stakeholders, including

- Certifying the conformance to some standard,
- Assessing the quality of the architecture,
- Identifying opportunities for improvement, and
- Improving communication between stakeholders.

An architecture review can be facilitated by adopting the processes defined by the *Software Architecture Review and Assessment (SARA) Report*, a practice that was developed by a group of IT/enterprise architects and published in 2002 (Kruchten, 2002). It applies a comprehensive evaluation model that covers a conceptual model of the IT architecture, reviews generic processes (workflows), and addresses typical activities that may take place in a review, such as identifying key stakeholders, architecturally significant requirements, standards and constraints, and risk assessment artifacts (Kruchten, 2002).

The evaluation should confirm that the architecture can answer these questions: What are the system objectives and requirements? What are the existing systems and constraints? How flexible/adaptable is the system? Does the system meet the needs of its users?

4.6.1 *What Are the System Objectives and Requirements?*

As stated previously, business objectives create the context in which the architecture exists. The system should, therefore, have as its goal delivering on the mission of the organization. The key characteristics for water and wastewater utilities are reliability, safety, rapid response to breaks, minimal service disruption and inconvenience because of construction, accurate billing, fair rates, compliance with regulations, and so forth. Overall IT systems that are sufficient to support these are broad and distributed across many functional areas.

A foundational item for evaluating the various systems is to create an inventory of user scenarios (i.e., use cases) from the point of view of system end users. Scenarios represent tasks relevant to different roles such as end user/customer, public relations, system administrator, maintainer, and developer. As discussed in Section 3.1, these can also be defined with use cases (see Section 2.2.1 of this chapter). In a broader, architectural realm, they are high-level descriptions of how each user type in the organization interacts with the IT infrastructure (i.e., provides input and receives output).

It is the entire set of intended interactions (scenarios) that describe the system requirements, which are needed to achieve the system objectives. Both requirements and objectives should be documented in a way that is understandable by nontechnical and technical users.

4.6.2 *What Are Existing Systems, Solutions, and Constraints?*

As described in Section 4.5.4.1, "Architectural Notations," the enterprise architect should adopt some kind of notation and diagram type to depict the current system at an appropriate level of detail for current analysis. Section 4.5.3, "Methodologies and Frameworks," outlines the SAAM model views that describe the system. This is a useful technique for evaluating current systems as each view provides a particular perspective on the software architecture.

Whatever the documentation approach, this evaluation step focuses on how prior assumptions and conditions helped form the current IT infrastructure as a way of gaining insight into potential future needs, disruptions, and opportunities. This step uses the current systems as a "case study" for what is, what may be impeding change, and what future systems might provide that would reduce current constraints (or deal with them more effectively). It is also an opportunity to rework archived documentation and obsolete designs that no longer reflect the as-is situation.

4.6.3 *How Flexible/Adaptable Is the System?*

In developing user scenarios, it is necessary to "develop task scenarios that illustrate...the kinds of changes which, it is anticipated, will be made to the system over time" (SEI, 2009b). As discussed in Section 4.4, "Themes in Water and Wastewater Utilities' Information Technology Architecture," anticipating change requires recognizing new demands and opportunities. A feature of good architecture is "modifiability," which is a reflection of the architecture's ability to change with less cost and disruption to ongoing activities. Analysis of scenarios that include change and future modifications will reveal which architecture(s) have more built-in adaptability and which additional factors emerge for further consideration.

4.6.4 *From Fitness to Task (Does It Meet the Needs of Its Users?)*

The combination of these evaluation tasks should result in a clear picture of a system at various levels and perspectives that can be applied to current and future planning activities. It may be necessary to conduct this work iteratively and incrementally and use the emerging documents as a basis for organization-wide discussions. Through illustration, review, and refinement, the architecture's true baseline (as-is state) will serve as a valuable reference. Future discussion and modeling of the target to-be architecture will have this foundation. The enterprise architect will have a set of tools with which to communicate across organizational boundaries.

In the role of technology communications, the architect must analyze and bridge the gaps between proposed solutions and organizational priorities. According to the article, "The Case for Enterprise Architects" (Nash, 2008), the successful enterprise architect "must not only connect with senior business managers, but with the rank and file IT staff as well." That the solutions and systems meet the needs of users is part of that analysis.

5.0 SOFTWARE DEVELOPMENT

Information technology groups in water and wastewater utilities engage in software development as part of their mission. Software development, a branch of software engineering, is the set of activities, specifically computer programming, that result in a software application. Software engineering is a young discipline compared to other engineering fields, with most of its growth occurring in the mid-to-late part of the last century. It is a dynamic, changing field with new tools and techniques created every year to add rigor and control the uncertainty that can exist in software development efforts. Software development is a rich blend of both art and science. This section presents a brief introduction to software development and describes established application development methods and tools to support development.

5.1 Introduction

During the past half century, city, state, and federal governments have invested in a massive buildup of IT to support the goal of better government. Although it is hard to imagine how any major U.S. city could operate without IT today, IT system capabilities often seem contrary to the intentions of governance.

A widely publicized report by Boston, Massachusetts-based The Standish Group International, Inc. (1995) examined the state of IT software development with stunning results. For instance, about 30% of projects were cancelled before they were completed. In addition, the budgets of about 50% of the projects doubled from their original estimate. Moreover, only 16% of the projects were completed on time and within budget and, typically, these implemented a fraction of their original specified features. Factors that contribute most to the success of a project are

(1) User involvement,
(2) Executive management support,
(3) Clear statement of requirements,
(4) Proper planning,
(5) Realistic expectations,
(6) Smaller project milestones,
(7) Competent staff,
(8) Ownership,
(9) Clear vision and objectives, and
(10) Hard-working, focused developers.

It is important to note that the top three factors relate to organizations that commissioned software development, whereas characteristics of the developer team are lower in the list.

Tools and techniques of software development have changed significantly in the years since the aforementioned Standish Group International report. The lesson learned from the past was that diligent management with a focus on well-defined, but evolving, goals is essential to software development success. Many improvements have been made, especially in understanding the dynamic nature of constructing a software application. This has led to improvements in methods and software technology. Modern methods and computer languages allow for rapid change, reusable components, and frequent feedback to gauge progress. Software technology and development tools now allow far more connectivity to data sources, targets, and other applications and more reliability and greater security. Software is also a lot easier to construct. These trends are expected to continue because of strong demand for capabilities that put the power of software creation into the hands of managers and decision makers and new programming techniques that increase the efficiency and performance of computer programmers.

5.2 Software Development Methods and Management

As discussed in Section 2.0 of this chapter, software projects are often compared to other types of engineering projects such as, for instance, building a bridge. This analogy is useful because constructing a system for managing information has many of the same steps as constructing a bridge, road, building, and so forth. Requirements are collected, designs are drawn up, models and prototypes are constructed for simulation, the system is constructed, and tests are performed to confirm that the final product meets design specifications.

Although the stepwise, civil engineering-like approach to software development is valuable, the analogy starts to fail when one considers the infinite number of alternatives in software development, the ease with which changes can be made, the difficulty of describing a priori what the software should do, and the lack of rigorous methods to guide development. Although these factors work against efficiency, they can be controlled. Availability of alternatives and flexibility are real advantages that developers and development managers can leverage to effectively control a lack of rigor and changing requirements.

5.2.1 *Development Life Cycle*

The typical steps of a software project are

(1) Requirements analysis,
(2) Design and architecture,
(3) Implementation/coding,
(4) Verification/testing and debugging,
(5) Installation/release, and
(6) Maintenance/improvement.

Each of these steps is critically important to the final product and, collectively, they represent a causal chain. The design depends on requirements, coding cannot be done until the design is complete, and so forth. Any software project will involve these steps and, as such, the ultimate goal in software development is to enable automation of each step based on the specification of requirements. Research into software tools based on "formal methods" is aimed at automating development of correct software. Formal methods synthesize running programs from fundamental software building-block and user-supplied specifications, with little or no requirement for a programmer. A big difference between the various software development methodologies is the way each methodology progresses through these steps. The traditional, non-iterative waterfall methodology proceeds sequentially, whereas modern Agile software development approaches iteratively cycle through these steps.

5.2.1.1 *Waterfall Method*

The waterfall model, or method, is a software methodology that steps sequentially through the development phases of analysis, design, implementation, testing, installation, and maintenance. The method assumes that progression to the next stage can only occur after the previous stage has been fully completed. Visually, the method can be pictured as a steady stream downward, that is, like a cascading waterfall as shown in Figure 6.18, which is a reproduction from a paper published by an early software development pioneer (Royce, 1970).

In "Managing the Development of Large Software Systems," Royce (1970) concluded that this approach does not work well. Despite this, the waterfall approach was widely adopted and continues to be a major reason for failure in software projects. However, the waterfall method did require a well-thought-out requirement analysis and design phase. This is because flaws in subsequent stages are expensive and a direct result of poor requirements and design. The waterfall method also emphasizes documentation upfront, which makes the project less dependent on individual developers.

Criticism of this approach is that it is impossible to complete and perfect each phase before moving to the next. Often, requirements and designs cannot be fully developed and new insights are learned during design, implementation, and testing. If clients change their requirements after the design is finished, the project has to go back to the requirements phase, thereby undoing a lot of design work that has already been done. These realities all point to the need for more iteration, a conclusion that even Royce (1970) noted in his paper (see Figure 6.19).

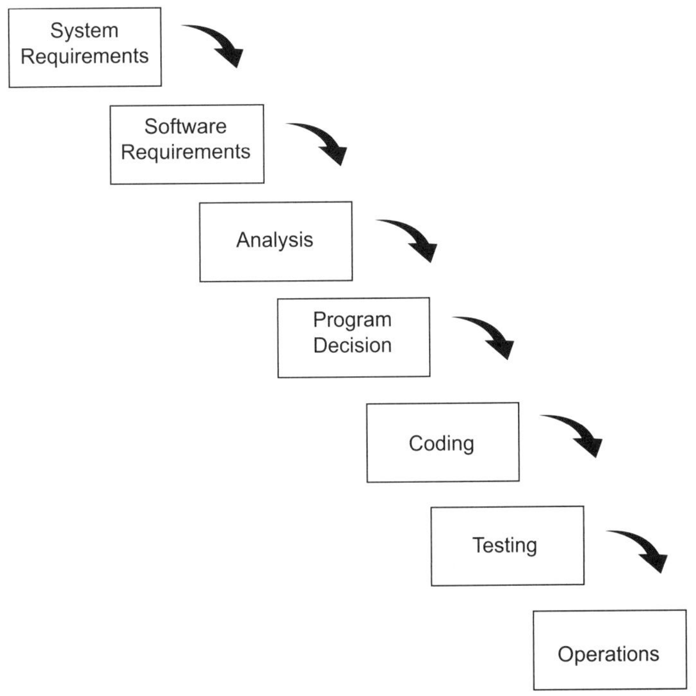

FIGURE 6.18 Winston Royce's depiction of the waterfall method (reproduced from "Managing the Development of Large Software Systems," Winston W. Royce—1970. © [1970] IEEE).

5.2.1.2 Iterative Methods

Experiences with different styles of software development have shown that it is extremely difficult to translate written requirements and specifications into high-quality software. Reasons for this difficulty are the inability to predict all future needs and a natural tendency to learn as developments progress. It is impossible to think of every detail in advance and, as capabilities are created and observed, we think of new ways to achieve the same goals or to enhance the original capability. These factors create uncertainty, which increases with undisciplined development, and cause "requirements churn" and "scope creep." Recognizing these realities, practitioners created Agile development methods (Martin, 2002), including the many variants such as Extreme Programming, SCRUM, and Dynamic Software Development Methodology, as a way to both control uncertainty and to take advantage of change.

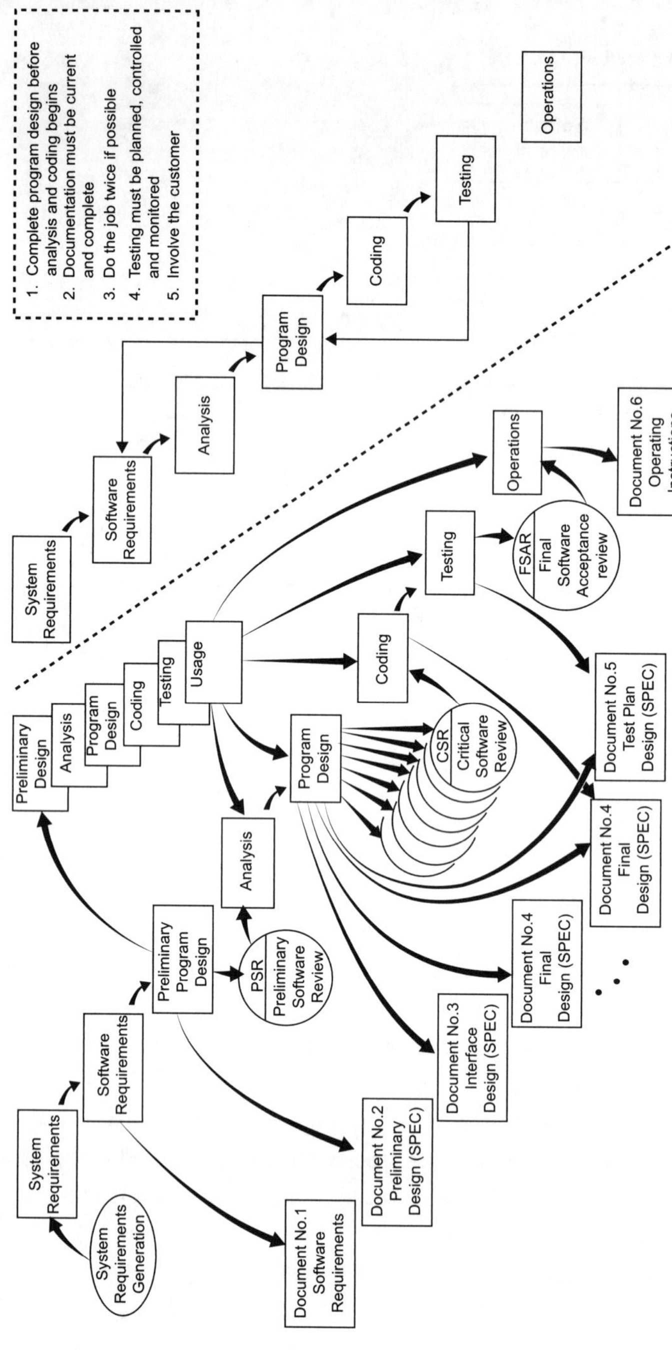

FIGURE 6.19 Winston Royce's conclusion showing the importance of iteration (reproduced from "Managing the Development of Large Software Systems," Winston W. Royce—1970. © [1970] IEEE).

Agile software development achieves results through the following best practices:

- *Customer involvement*—Customer involvement and continuous feedback is a critical factor to a successful project. In Agile projects, the customer is involved throughout the project life cycle, from creating the vision to planning iterations, defining and adjusting requirements, and testing and validation.
- *Standards alignment*—Agile encourages adoption of standards that enable applications to be easily understood and maintained. Developers follow a well-defined and proven set of standards and guidelines that guarantee applications will have a common style. Adherence to standards achieves quality and consistency and facilitates collaborative development and refactoring.
- *Small releases*—Agile planning breaks projects up into short iterations or release cycles, each cycle containing design, construct, test, validate, and integrate phases. Agile planning with small-release cycles gives managers control over the project by allowing them to change direction when needed and to eliminate nonproductive efforts early.
- *Test-driven development*—Automated tests of features are created before features are developed. Tests are continuously enhanced as new features are added and customers have access to these tests to monitor development performance. This ensures quality and software that always works as specified.
- *Refactoring*—By continually reassessing features through tests and focusing on both current and existing capabilities, developers ensure that the design and structure of an application is of the highest quality. Refactoring allows developers to improve the design of an application and maintains it when new features are added.
- *Continuous integration*—Continuous integration avoids problems of application assimilation (i.e., integration or linking component modules together) that occur with traditional methods that leave this important task to the end of the project. By adopting this approach, the application will always be in a functional state and the end user of the software application is able to validate features at each step during the project.

Agile software has other useful practices such as pair programming that is suitable for larger development efforts, which requires two programmers: one to write code and the other to review code as it is written. Agile planning methods require specification of "feature points" corresponding to the feature to be developed, mapping these points over time to an amount of effort required to create the feature and, based on this information, providing management with hard metrics that show how quickly the application is being developed and when delivery can be expected. This planning practice alone represents a significant advancement over other methods, which try to achieve deadlines by dictate or with great difficulty.

Agile methods have gained popularity because they are effective at accommodating the change and uncertainty that typify software development projects. Agile methods promote collaboration with stakeholders, encourage teamwork and mutual accountability, and deliver tested software, resulting in higher-quality and software features that better align with business needs.

5.3 Software Development Tools

Software development tools are as rich and varied as the types of software applications from which these tools are constructed. Software development occurs in a computer environment comprising programming code editors, debuggers, code- or document-version control systems, documentation creation systems, and machine or human-interface development tools. At one end of the development environment spectrum, there are simple text editors and command-line code/buildtools. At the other end of the spectrum are fully featured integrated development environments (IDEs) that have all the necessary tools built into one package.

Software applications typically are built for a specific computing platform, which is the target executing environment for the software application. There are two prevalent target platforms: (1) Microsoft Windows® and variants (e.g., Windows® CE, etc.), and (2) Java (Oracle, Redwood Shores, California). These two platforms have dominated for many reasons, among them being the technical difficulty in creating a platform, the importance of a large community of developers, the size of the market, and various technical capabilities. Integrated development environments, both commercial and open-source, have developed on these two platforms. Some IDEs are specialized by application vendors for their software (e.g., Oracle) and others are unique to a class of software application (e.g., industrial automation). The two dominant IDEs are Microsoft's Visual Studio line (www.microsoft.com/visualstudio) and Eclipse (www.eclipse.org).

Organizations either enforce a standard relating to platform or they require conformance to standard practices without imposing a platform standard. Decisions related to computing platform should be made based on criteria discussed within the company, including both IT and non-IT groups. The choice has significant effect on the way software will be developed, the vendors available to work with, and the level of support and cost.

5.3.1 *Structured Methods for Software Development*

The six steps described in Section 2.2.1, "User Requirements for Software Procurement," involve two equally important groups:

- Managers and other stakeholders who are experts in understanding business motivations driving creation of a software application, but not necessarily experts in IT; and
- Computer engineers and programmers who are experts in IT, but not necessarily experts in business motivations driving the creation of a software application.

When these two groups come together to create software, the differences in experience and perspective can create a gap that defeats the common goal. This gap is recognized by the software development community, but there is no single, agreed upon solution. Most successful efforts apply methods that clarify business processes and technical activities, which streamlines the six steps in software development, described in Section 5.2.1.

Two software industry efforts have resulted in holistic views of how systems can be built:

- Open-source: OMG's MDA and UML and
- Commercial: IBM's rational unified process (RUP).

Both efforts maintain that the methods apply to managing change and the many types of systems that support the business, including software. Here, the focus is on the use of these methods for streamlining software development, which has resulted in many successes. Use of these methods, however, entails a significant commitment and considerable learning to implement properly. Although they are not essential to a software development effort, they do provide many useful tools one can choose from to advance a development project.

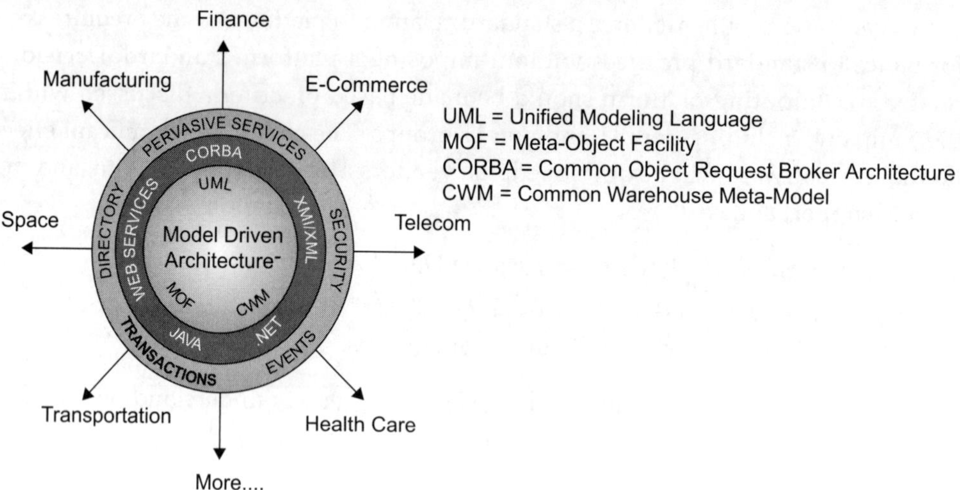

Figure 6.20 Object Management Group's (OMG) Model-Driven Architecture (MDA) (MDA diagram reprinted with permission. Object Management Group, Inc. © OMG, 2009).

The Object Management Group developed MDA (see Figure 6.20) over many years, slowly evolving graphical tools to support object-oriented development and software technologies that enabled rapid integration to reach their goal of delivering open-enterprise integration standards (visit www.omg.org/mda for more information). One widely used modeling tool is UML, the de facto standard for modeling software systems including object models, data models, workflow models, and models of behavior that implements software capabilities. Model-driven architecture separates business and software application logic from the ultimate platform to be used. Separating business functionality from technology implementation, while providing mechanisms to keep them in synch, enables each to evolve somewhat independently, thus resolving the business-technology gap.

The acquisition of RUP by IBM occurred when the company purchased Rational Software Corporation, a company whose mission was to provide open and standardized software tools and best practices. Rational Software brought together several experts in software development and created a process that they felt aligned business objectives with software technology. IBM'S RUP is a software engineering process that imposes a discipline on development specifically to bridge the business-technology gap. Rational Unified Process (see Figure 6.21) provides software tools

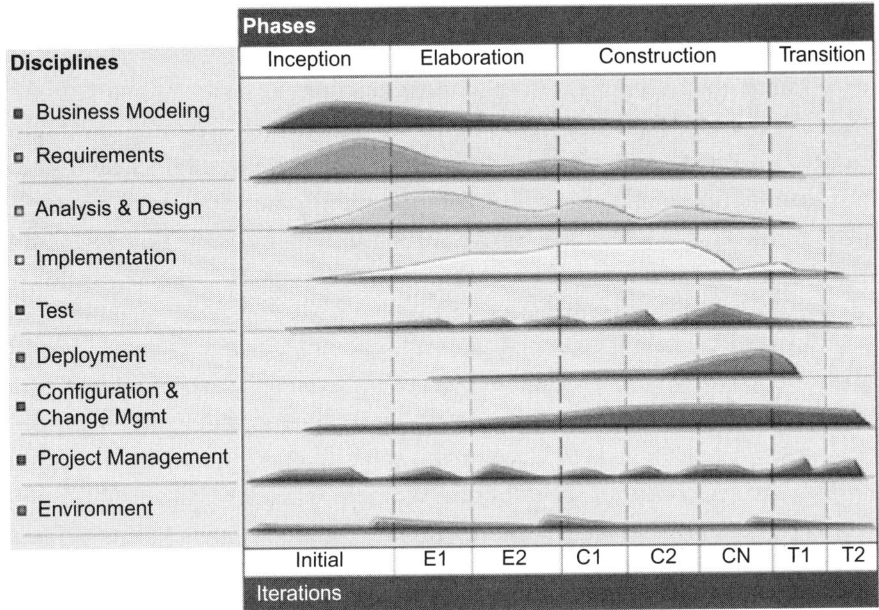

FIGURE 6.21 IBM's RUP (IBM, 2009).

for modeling software (including UML, the standard for which is maintained by OMG), is configurable in that it can be modified to suit different styles or incorporate other methods of development, and captures the following six best practices of software development:

- Develop software iteratively,
- Manage requirements,
- Use component-based architectures,
- Visually model software,
- Verify software quality, and
- Control changes to software.

6.0 SOFTWARE PROCUREMENT

As described in Chapter 2, there are many COTS software products that address a wide variety of business needs for water and wastewater utilities. Such software products range from GIS, LIMS, and mathematical models to billing applications or business/financial/human resources systems (like Enterprise Resource Planning and human resources packages).

Purchasing a COTS product with standard functionality is often preferable to customized software development. Using a standard version of a software product that has broad acceptance in the marketplace is much safer than creating a "one-of-a-kind" system. Therefore, before starting customized software development, it is important to justify such a need. Users sometimes believe that their business needs are "special" and different from the needs that users at similar organizations have. It is important to analyze this and determine if the nature of the underlying business process is, in fact, different and special. Sometimes, it is better to make changes to underlying business practices to find a COTS solution that will support it.

This section focuses on procurement of major IT "components" such as software applications, major hardware components, and the services related to their acquisition and release, including COTS solutions. It does not address consumables (i.e., supplies such as printer ink cartridges, compact disks, etc.) and other related components (i.e., cables, connectors, wireless communication cards, etc.), which are frequently charged to an IT budget. For the purpose of this section, there are no differences in approach to procurement when done through either capital or operations and maintenance (O&M) budgets. Budgets for capital and O&M are based on different criteria, such as the source of funding; this section, however, pertains to IT procurement, not funding. It is also written from the perspective of the utility making the purchase and not from the perspective of a vendor. Every utility should understand that a software vendor might not treat their sales opportunity the same if the funding source being used is foreign to their normal mode of business and may force them to "sign up" for contractual obligations not in their normal mode of delivery, payment, and support. The key point is that IT is an important asset that needs to be planned and procured, like all other major purchases, and should be part of a utility's overall planning processes.

Whereas the work involved in defining, obtaining, and launching IT systems is tasked to the "technical" groups of wastewater utilities, procurement is often the purvey of the purchasing group. However, this is often a source of failure in the procurement of IT in that the methods used to make most wastewater purchases do not apply well to IT systems. Indeed, IT is so different from pipe, concrete, pumps, and trucks that the terms and conditions, legal requirements, specifications, general conditions, and so forth of the purchasing process need to be tailored for this unique acquisition.

The fundamental question to ask is, "When you buy software, what are you really buying?" This simple question has many different answers that need to be clearly understood before making procurements. For example, with certain pieces of software, you are not actually buying a product but merely the license to use the product. Often, the actual product remains the property of the software company or the developer; unless you specify that you want ownership of the end product.

6.1 Laying the Foundation for Sound Information Technology Procurement

6.1.1 *Information Technology Strategic Planning*

In terms of IT investments, management staff of utilities must anticipate future needs, allocate limited resources among competing demands, and manage changes in the organization and its processes associated with adopting new technologies. In addition, they must take necessary actions to contain costs, increase efficiencies, and put IT in long-term compliance with regulations. This complex business climate demands improvement of IT use. Not doing so could mean continuing current practices with the procurement of IT disconnected from the rest of the business operations. To develop integrated strategic and tactical plans requires an IT master plan (ITMP). An ITMP delineates a comprehensive, consolidated effort to establish, maintain, and evolve effective information systems throughout the enterprise.

An ITMP organizes and coordinates the current IT investment and creates strategic and tactical plans for the acquisition and implementation of future technology purchases. The following are goals of an ITMP:

- Development of an integrated information system to allow functions, users, and systems to share needed information, reduce redundancy, and improve the integrity of data;
- Improvement of staff productivity through better access to shared data, creating links between systems that need to post or download data, and providing access for all authorized persons;
- Enhanced coordination and communications by providing the ability to exchange timely, relevant data in compatible formats;
- Definition of coordinated methods, procedures, and standards for program planning and evaluation of all information systems; and
- Improvement of analysis, decision-making, and reporting capabilities.

In the process of developing an ITMP, there are methodologies and frameworks that facilitate an understanding of where the utility is now with its information systems and technologies, where it wants to be, and what its expectations are of how to get there.

The best methodologies use an organization-wide (i.e., enterprise) view, which focuses on how the enterprise can use IT to support its mission, vision, and business objectives. Because business processes determine information requirements, the enterprise's current business processes are typically assessed and an understanding of the collective future vision is gained (i.e., the enterprise's business plans, goals, and objectives). The basis of an ITMP comes from mapping the differences between current operations and the way the utility envisions operating in the future. Based on these differences, technology solutions to progress from the current to future state are defined, and the supporting background to justify their procurement is created.

As individual projects described in a ITMP are undertaken, sufficient care must be taken to do a complete scope development for them. Information technology master plans can have varying levels of scope included for each project, ranging from a brief overview to increasingly more detailed definitions of requirements, expectations, interfaces, goals, platforms, and so forth. As cited in the introduction of this chapter and in Section 5.1, poor scope development early on in an IT project is often the cause of budget overruns and schedule changes. Spending ample time up front to get the scope complete and agreed to, with management buy-in and sign-off, is vital to achieving desired end results. Work plans with task breakdowns, staffing allocations, budgets, and schedules all need to be in place before a project is started. These documents can then be the tracking tools to follow IT projects through to their completion. Any changes that are desired or become available during the project's life can be discussed and weighed against scope, schedule, budget, and staffing effects, with eventual decisions made in light of a complete set of criteria with which to perform an analysis.

Techniques of information engineering can be used to match the information needs of management and staff with the data being provided by the procured software systems. This allows the generation of a model for the utility that contains the data/information required to support business requirements. One definition of information engineering is as follows:

> An integrated set of methodologies and products used to guide and develop information processing within an organization. It starts with enterprise-wide strategic planning and ends with running applications. (PC Magazine Encyclopedia)

Alternatively, Wikipedia (2009c) offers the following:

> Information engineering methodology is an architectural approach to planning, analyzing, designing, and implementing applications within an enterprise. It aims to enable an enterprise to improve the management of its resources, including capital, people and information systems, to support the achievement of its business vision. It is defined as: "An integrated and evolutionary set of tasks and techniques that enhance business communication throughout an enterprise enabling it to develop people, procedures and systems to achieve its vision." Information engineering has many purposes, including organization planning, business re-engineering, application development, information systems planning and systems re-engineering.

Both of these definitions demonstrate that using techniques that clearly define the scope of IT procurement relative to the overall needs of an organization and its enterprise architecture is a critical step in successful procurement of every part of a utility's complete IT.

By capturing information requirements in this way, managers can more readily understand and plan for the use of information in decision-making processes. Additionally, outmoded data structures and data management processes can be redesigned to allow for more effective use and maintenance of new systems.

6.1.2 *Establishing Information Technology Priorities*

Once information system needs have been identified from a master plan, priorities for their procurement can be established. These are balanced with the utility's overall needs so that the most needed procurements are made first. Priorities are typically set to accomplish three major categories of IT infrastructure: technical platform, core applications, and decision support tools. The technical platform refers to all IT procurements that build the "foundation" for all IT (i.e., the desktops, handhelds, servers, networks, interfaces, communications, etc, that make up the technical backbone). Core applications refer to any software that performs a core business function of the utility (i.e., management, operations, engineering, design, maintenance, mapping, procurement, customer information, billing, lab, etc.). Finally, decision support refers to those IT procurements that step beyond the basic "off-the-shelf" purchases and address enhanced applications such as dashboards, portals, business intelligence, unique integrations of applications, and more. Whatever a utility chooses as their definition of these basic categories, it becomes the utility's responsibility to establish its priorities for procurement in each of these areas, as they support its mission, vision, and objectives. The technical platform is those components that make up the hardware, networks, and communications to support the overall IT architecture. Core applications are those IT software products that support the basic business functions or operations, maintenance, and management. Decision-support tools include all IT applications and services that leverage a combination of applications and draw intelligence from multiple sources to achieve new efficiencies.

There is no standard analysis to apply for setting the priority of IT purchases. Depending on where a utility sits in the continuum of IT procurement, such as whether they are sophisticated users of IT or just beginning to build their enterprise architecture, their business needs determine their IT needs priorities. Determining the return on investment for IT purchases is an excellent way to assess these purchases vs other needed investments. Often, an IT investment helps leverage a previous expenditure and may extend or supplement its life or usefulness, thereby avoiding a more expensive capital expenditure with a less-expensive IT purchase.

6.1.3 *Purchasing Policies as Part of Information Technology Governance*

Every utility must buy in accordance with good governance. Governance must exist in a recognized structure, with executive support and appropriate membership from the organization. In terms of IT procurement, some purchasing policies might need to be adapted to account for the unique nature of the purchase. Some utilities are creating their own governance models, whereas others are adopting models such as Information Technology Infrastructure Library (ITIL). Although ITIL can influence the purchasing process, it does not define it. Standards that should exist for IT must include standards for defining and controlling procurement.

6.2 Information Technology Acquisitions as Investments

6.2.1 *Investment Evaluation and Priority-Setting Policies*

Best practices are those operating rules and procedures of an enterprise that allow it to achieve operational excellence, which includes having least costs, high productivity and quality, a satisfied staff, and recognized leadership in the industry. With pressure from regulators, customers, and the courts, the utility needs to incorporate the best practices of its industry into its business processes.

After properly identifying IT needs, there must be practices for prioritizing those needs. Business processes used for these evaluations can be compared to the following best-practice criteria being used today:

- *Meeting regulatory requirements*—This practice ensures that appropriate attention is paid to water quality, public health, and environmental stewardship.

Being effective at meeting regulations in the information age requires that a utility provide personnel with adequate technological tools to respond rapidly to situations that could compromise compliance.

- *Enterprise asset management*—This practice seeks to manage an enterprise's assets to optimize their use, thereby maximizing the return on investment in the assets. It focuses on three critical business functions within an organization: asset identification and evaluation, maintenance management, and work-order response. Enterprise asset management begins with the identification of assets and their financial valuation. This allows an organization to assign accurate values to its structures, facilities, equipment, customers, information, systems, employees, and so forth through a variety of financial analyses. This valuation provides the basis for asset capitalization and asset maintenance life-cycle planning. Good maintenance management strategies and efficient management of work orders provide the most effective level of maintenance services, maximize the life of critical and costly assets, and minimize the overall cost of maintenance.

Performance-centered business management recognizes that effectiveness and efficiency are the most important factors guiding all activities of an enterprise. An organization that is striving to improve must first assure its stakeholders that it can operate effectively and then work to achieve the best operational efficiency feasible. These challenges demand that business management principles be applied to the operation of a utility, including the development of a business strategy and the establishment of links between information systems and the business strategy. To link information systems implementations to the strategy that is driving the business, the following need to be clearly defined:

- Business mission statement,
- Business objectives,
- Business critical success factors,
- Business performance measures, and
- Utility-wide business data structures.

Once these factors are defined, mechanisms need to be established to measure the effectiveness of work processes and business decisions and then compare the existing effectiveness to performance targets. This often requires the ability to track the total costs of performing a set of activities (e.g., purchasing). Activity-based accounting can assist decision making in this area, but performance-centered business management requires much more than just activity-based accounting.

6.2.2 *Building Business Cases*

Once consistent and sustainable operational effectiveness has been achieved, a business can focus on achieving higher levels of operational efficiency. Another set of mechanisms can then be instituted through an operational efficiency improvement program that plans and implements the means by which an organization is to achieve its improved operational efficiency targets. As part of any operational efficiency improvement program, it is essential that management and employees be provided with the information required to examine, critique, and enhance existing facilities, equipment, and work practices within the organization, and to implement and add new ones. Therefore, there is a two-part role for information services within an organization that has instituted an operational efficiency improvement program:

(1) Information services, as part of the improvement program, must work to set new performance targets and then manage to achieve those targets and
(2) Information services must provide the information that everyone else in the organization needs to achieve their own performance targets.

6.3 Managing the Information Technology Procurement Process

6.3.1 *Using a Life-Cycle Information Technology Approach*

Considering the speed at which technology changes, and the many consolidations and business transformations that exist for vendors of IT, many may view life-cycle planning as too difficult a task; however, IT budgeting needs this type of view to be realistic. Too often, the cost of IT acquisition is viewed as just the initial purchase of a technical commodity without consideration of all associated costs that will be incurred throughout its life. Therefore, when asking for IT funding, it is appropriate to consider all components required to obtain, launch, use, leverage, support, and operate each purchase. Each individual purchase must be considered in the complete environment in which it is going to operate, all the way up to the enterprise level, and even considering connections to entities external to the organization.

Life-cycle planning is intended to provide coordination and to allow for IT to be treated as every other asset a utility manages and uses. *Life cycle* is defined as the useful financial life of an item, and should capture the number of years a piece of hardware or software is intended to be kept. At the end of a life cycle, the item is either no longer suited for its intended purpose or maintenance and support have grown to the extent that it is cheaper to replace it than keep it or new requirements have necessitated its replacement to meet needs. This can be challenging and requires some assumptions that need to be stated and documented. For example, changes in technology might obsolete an application or a piece of hardware before its expected life. If the stated requirement is to keep an application running, then the budget to handle the situation would be justified.

Life cycles for IT are not static, but vary depending on the technology being used and the rate of change inherent in the industry for its type. The challenge for budgeting is that IT does not operate on the same life cycles as pipe or other assets. Periods of less than 7 years, and as low as 2 or 3 years, are common for IT life cycles. Life-cycle planning allows for a common perspective for the planning and sizing of IT investments.

As utilities are increasingly asked to do more with less funding, the need to automate and computerize will continue to grow. Information technology is one of the best hopes for improving productivity. The life cycle must evaluate the trade-offs inherent in technology and reach financial justification for IT purchases. The process must recognize the inevitability of change and have a long-term perspective. Most importantly, the process must consider all related costs, such as the services that might accompany a purchase for installation or maintenance (sometimes 2 or 3 times the cost of the product itself), peripherals, infrastructure (e.g., wiring or electrical), and ongoing annual fees.

When using the life-cycle approach, future computing trends must be considered. The change to cloud computing, or thin client technology, may offer a superior computing environment over the long run by allowing the purchase of lower-cost computers with less onboard components rather than buying more expensive models and trying to make them last longer.

Life-cycle planning for IT is not easy, as predicting changes and future directions for IT will always be difficult. The biggest danger is underestimating the complete cost of information systems and underestimating the accelerating rate of change inherent in ITs. A good planning process will temper optimism with reality, improve coordination, identify all components of costs, and help ensure that the returns on investments are well-founded and achievable.

6.3.2 *Establishing Organizational Structure for Information Technology Procurement*

Once the project has been defined by the appropriate planning process, it needs to be managed to successful implementation and operation. The same organizational structure that oversees the IT architecture should assist with management of IT procurement. A centralized group with oversight of the acquisition and implementation of all IT within an organization is the best approach. This group ensures that any new system is compatible with the enterprise's technology vision and architecture and that the investments will facilitate the enterprise in meeting its stated objectives. The group must also develop and maintain short- and long-term information system plans.

Information technology governance is essential to support the continued mission of business operations. Effective and efficient governance is also critical to the success of IT procurement. Information technology governance organizations focus on the value of IT investments and, as such, understand opportunities in the procurement process to transform the way IT helps do business.

Information technology governance is about the way in which leadership accomplishes the delivery of mission-critical business capability. Information technology governance is concerned with strategic alignment between the goals and objectives of the business and the use of its resources to effectively achieve the desired results. Good governance can change the organization's view of IT as purely a cost center and show the true value IT adds to the utility. Clear governance policies help translate IT activities into measurements of business value.

The group tasked with IT governance will need documented processes and practices that manage the collection of all IT applications like a portfolio of investments. Factors such as risk, desired improvements, best practices, and project management are all under the governance of this group. There are some known structures for IT governance, and they can be adopted for a utility in part or in whole. They address the total governance of IT and not just the procurement processes, including the overall standards needed to coordinate the IT infrastructure.

6.3.3 *Defining Terms and Conditions for Information Technology Procurement*

One of the first areas to look for changes in the IT procurement process is to examine the terms and conditions used in the contracting process. Most utilities use a standard specification process, which includes a section format, beginning with general conditions and specific sections for specifying the totality of components being purchased. These general and specific components are then wrapped in terms and conditions ("Ts" and "Cs") and other "boiler plate" language that define the purchasing process and requirements. Although these words are appropriate for most purchases that a utility might make, they are not good for purchasing IT and need to be revised to accommodate the unique demands of IT.

There should be a glossary included in the purchasing document to clearly define terms and their intended meaning. No matter how well the terms and conditions are defined, most companies that provide IT components and services have their own terms and conditions they will expect the purchaser to accept; these typically comprise lengthy documents that are not open for negotiation.

Normal payment practices probably will not apply to the purchase of IT services and applications. Whereas it might make sense to pay based on a percentage of work completed for yards of concrete poured or feet of pipe installed, it becomes a difficult process to determine the "percentage completed" of an IT application that truly only works after it is completely programmed, installed, tested, and operating.

There are free sources for help on procurement terms and conditions from federal government agencies and many educational institutions; these can be found using simple searches on the Internet and from IT standards organizations, the latter of which might charge a fee for their published procurement guidance documents and methodologies.

6.3.4 *Evaluation Factors*

Purchasing documents should have an agreed-to set of evaluation factors and protocols for using them when reviewing bids and proposals. Although evaluation factors can vary widely depending on the IT investment being made, they must always provide a way to ensure that the procurement meets stated business needs and objectives. These might include evaluations of ease of use, references from existing users, timeliness of previous engagements, documentation, training, flexibility, communications, interface capability, support, history of bugs and fixes, and other such categories. As the team develops these evaluation factors, they must decide whether or not they are "musts" or just "desirables." Weightings are then placed on factors to appropriately balance the evaluation process in the right direction and not let minor categories unduly influence proper scoring and selection of the best offer.

6.3.5 *Formalizing the Contract*

The formal document must include payment mechanisms, progress measures, responsibilities, completion criteria, definitions of (substantial and final) completion, renewals, extensions, dispute resolution, support, and payments.

6.4 Sample Information Technology Sourcing Scenarios

There are many ways to procure IT services and components; this section touches briefly on some of the more prominent methods used. Regardless, the method chosen must support the end objectives of the business.

A utility can choose to hire an external entity to provide IT resources and have them perform in their facilities under their own supervision and direction. This method can have more external resources in project management roles and supervision. This can extend to having an external resource performing as the project management office for IT application construction and operation.

There are also several methods for obtaining software applications that range from owning the application completely to purchasing a "hosted" version, known as Software as a Service (SAAS).

As with all IT terms, the starting point for discussing a subject is defining it. As previously mentioned in this chapter, SAAS has become a valuable IT resource. In terms of a utility's overall application portfolio, SAAS is a way to launch software by hosting it remotely and providing users access over the Internet. It often uses a single code base for all users in what is called a *multi-tenant architecture*, although this is not the only architecture offered by these types of vendors. The code is typically not customizable for an individual user, but is the same for all. This has some advantages in the management of IT, which include faster implementation of an application, easier access to current technology, significantly lower cost, and fewer errors in the software. However, SAAS is still software and, as such, shares the same problems as other software products.

An example of how SAAS might be used is for Web conferencing. A utility would not have to invest in the purchase of all of the components to do this activity, but could buy a service to accomplish this without having to extend its existing IT infrastructure.

7.0 REFERENCES

Ambler, S. W. (2003) *Architecture Envisioning: An Agile Best Practice.* http://www.agilemodeling.com/essays/initialArchitectureModeling.htm (accessed May 2010).

Armour, F.; Miller, G. (2001) *Advanced Use Case Modeling;* Addison-Wesley: Reading, Massachusetts.

Artisan Software Tools (2008) Architecture Web Site. http://www.artisansoftwaretools.com/capabilities/lifecycle-support/architecture/ (accessed Aug 2009).

Beck, A. R.; Fu, G.; Cohn, A. G.; Bennett, B.; Stell, J. G. (2007) *A Framework for Utility Data Integration in the UK.* http://www.comp.leeds.ac.uk/mtu/pubs/UDMS2007ARB.pdf. School of Computing, University of Leeds, U.K.

Boehm, B. W. (1981) *Software Engineering Economics;* Prentice Hall: Englewood Cliffs, New Jersey.

BSDGlobal (2007) Business and Sustainable Development: A Global Guide. *Green Procurement*. http://www.bsdglobal.com/tools/bt_green_pro.asp (accessed Aug 2009).

Carnegie Mellon Software Engineering Institute (SEI) (2009a) *Software Architecture for Software-Intensive Systems*. http://www.sei.cmu.edu/architecture/sara.html (accessed Aug 2009).

Carnegie Mellon Software Engineering Institute (SEI) (2009b) *Scenario-Based Analysis of Software Architecture*. http://www.sei.cmu.edu/architecture/scenario_paper/ieee-sw1.htm (accessed Aug 2009).

Cockburn, A. (2001) *Writing Effective Use Cases;* Addison-Wesley: Reading, Massachusetts.

Hammer, M. (1990) Reengineering Work: Don't Automate, Obliterate. *Harvard Business Rev.*, Jul/Aug, 104–112.

Hammer, M.; Campy, J. (1994) *Reengineering the Corporation;* Harper Collins: New York.

Hilliard, R. (2000) *IEEE-Std-1471–2000 Recommended Practice for Architectural Description of Software-Intensive Systems*. http://www.enterprise-architecture.info/Images/Documents/IEEE%201471–2000.pdf (accessed May 2010).

IBM Rational Unified Process data sheet. ftp://ftp.software.ibm.com/software/rational/web/datasheets/RUP_DS.pdf (accessed May 2010).

KPMG LLP. (1997) *IT Project Management Survey;* KPMG LLP.: Toronto, Canada.

Kruchten, P. (1995) *The 4+1 View Model of Architecture*. http://www.cs.umb.edu/~jxs/courses/2007/680/papers/kruchten.pdf (accessed Aug 2009).

Kruchten, P., Ed. (2002) *Software Architecture Review and Assessment (SARA) Report, Version 1.0*. http://philippe.kruchten.com/architecture/SARAv1.pdf (accessed Aug 2009).

Leganza, G.; Brown, A. (2004) *EA Frameworks Book Provides an Overview of Leading Models*. http://www.enterprisearchitecture.info/Images/Documents/Jaap%20Schekkerman.pdf (accessed Aug 2009).

Marshall, C. (1999) *Enterprise Modeling With UML: Designing Successful Software through Business Analysis;* Addison-Wesley: Reading, Massachusetts.

Martin, R. C. (2002) *Agile Software Development: Principles, Patterns, and Practices*. Prentice Hall: Upper Saddle River, New Jersey.

Nash, K. (2008) *The Case For Enterprise Architects*, CIO Magazine article. www.cio.com/article/472429/The_Case_for_Enterprise_Architects?page=4 (accessed Aug 2009).

Object Management Group (2009) Model Driven Architecture Home Page. www.omg.org/mda (accessed May 2010).

Royce, W. A. (1970) Managing the Development of Large Software Systems. *Proceedings of the Western Electronic Show and Convention; Los Angeles, California, Aug 25–28; Institute of Electrical and Electronic Engineers*: New York.

Schwaber, C.; Gerush, M. (2008) *The Forrester Wave Requirements Management*, Q2 2008. Forrester Research, Inc.: Cambridge, Massachusetts.

Schwaber, C. (2006) *The Root of the Problem: Poor Requirements;* Forrester Research, Inc.: Cambridge, Massachusetts.

Sinur, J.; Hill, J. B. (2009) *Selection Criteria Details for Business Process Management Suites;* Gartner Research Report; Gartner, Inc.: Stamford, Connecticut.

Standish Group International, Inc. (1995) *CHAOS Report*. http://net.educause.edu/ir/library/pdf/NCP08083B.pdf (accessed Aug 2009).

Temnenco, V. (2007) *TOGAF or not TOGAF: Extending Enterprise Architecture Beyond RUP*. http://www.ibm.com/developerworks/rational/library/jan07/temnenco/index.html (accessed May 2010). IBM Corporation: Armonk, New York.

The Open Group (2004) *Business Executive's Guide to IT Architecture*. http://www.opengroup.org/onlinepubs/7699959699/toc.pdf (accessed Aug 2009).

The Open Group (2009) *The Open Group Architecture Framework—Version 9;* The Open Group: San Francisco, California.

U.S. Department of Defense (2007) *Green Procurement Strategy*. http://www.ofee.gov/gp/gppstrat.pdf (accessed Aug 2009).

Von Halle, B. (2002) *Business Rules Applied: Building Better Systems Using the Business Rules Approach*. Wiley & Sons: New York.

Watts, H. (1989) *Managing the Software Process;* Addison-Wesley: Reading, Massachusetts.

Wikipedia (2009a) *Client Server*. http://en.wikipedia.org/wiki/Client-server (accessed Aug 2009).

Wikipedia (2009b) *Business Process Modeling Notation (BPMN)*. http://en.wikipedia.org/wiki/BPMN (accessed May 2010).

Wikipedia (2009c) *Information Engineering*. http://en.wikipedia.org/wiki/Information_engineering (accessed Aug 2009).

8.0 SUGGESTED READINGS

Brooks, F. P., Jr. (1995) *The Mythical Man-Month: Essays on Software Engineering*, 20th ed.; Addison-Wesley: Reading, Massachusetts.

Business Process Management Initiative and the Object Management Group™ (2008) Home Page. www.bpmi.org (accessed Aug 2009).

Business Process Management Institute (2008) Home Page. www.bpminstitute.org (accessed Aug 2009).

Business Rules Community, Frequently Asked Questions Web Page. http://www.brcommunity.com/faqs.php (accessed Aug 2009).

Carnegie Mellon Software Engineering Institute (2006) CMMI for Development v1.2. http://www.sei.cmu.edu/publications/documents/06.reports/06tr008.html (accessed Aug 2009).

Effective Utility Management Collaborative Effort (2009) *The Resource Toolbox—The Ten Attributes of Effective Utility Management*. http://www.watereum.org/10attrib.php (accessed Aug 2009).

IBM Corporation, Rational Software Home Page. www.ibm.com/rational (accessed Aug 2009).

Lassing, N.; Bengtsson, P.; Bosch, J.; Rijsenbrij, D.; van Vliet, H. (2000) *Modifiability through Architecture Analysis*. http://www.cs.vu.nl/~hans/publications/y2000/LAC2000.pdf (accessed Aug 2009).

Martin, R. C. (2002) *Agile Software Development: Principles, Patterns, and Practices*; Prentice Hall: Upper Saddle River, New Jersey.

McCarthy, J. (1995) *Dynamics of Software Development*; Microsoft Press: Redmond, Washington.

PCMAG.com Encyclopedia (1981–2009) Definition of information engineering. http://www.pcmag.com/encyclopedia_term/0%2C2542%2Ct%3Dinformation+engineering&i%3D44941%2C00.asp (accessed Aug 2009).

Requirements Development e-zine Home Page. http://www.requirementsdevelopment.com/ (accessed Aug 2009).

Shore, J.; Warden, S. (2008) *The Art of Agile Development;* O'Reilly Media, Inc.: Sebastopol, California.

Sparx Systems Enterprise Architect (Resources Section). http://www.sparxsystems.com.au/resources/index.html (accessed Mar 2010).

Supply Chain Council, Inc., Home Page. www.supply-chain.org (accessed Aug 2009).

The Open Group Home Page. http://www.opengroup.org (accessed Aug 2009)

Uhrick, S.; Feinberg, D. (1997) Integrating GIS With Utility Information Management Systems. *Proceedings of Geospatial Information & Technology Association 20th Annual Conference and Exhibition*; Nashville, Tennessee, Mar 23–26. http://www.gisdevelopment.net/proceedings/gita/1997/bepm/bepm20a.asp (accessed Aug 2009).

Value-Chain Group Home Page. www.value-chain.org (accessed Aug 2009).

Vitasovic, Z. (2005) *Decision Support Systems for Wastewater Facilities Management;* Report No. 00-CTS-7; Water Environment Research Foundation: Alexandria, Virginia.

Weigers, K. (2009) Process Impact Resource Page. http://www.processimpact.com/goodies.shtml (accessed Aug 2009).

Chapter 7

Information Technology Security

(*continued*)

1.0 THE NEED FOR INFORMATION TECHNOLOGY SECURITY IN WATER AND WASTEWATER UTILITIES

Information technology (IT) security, more specifically, industrial cyber security (henceforth referred to as *cyber security*), deals with the protection of enterprise information systems from external or internal attacks in industrial environments, including water and wastewater facilities. As such, it includes protection of people, production, assets, data, and deliberate intrusions into the control system infrastructure. The field of traditional IT security is well covered in other publications, and the IT departments of most organizations are familiar with the tools and methods available to protect the business enterprise systems. The terrorist attacks of September 11, 2001, have increased awareness of the threats and vulnerabilities that face utilities in the United States and elsewhere. A typical wastewater utility relies on automatic process control systems, including supervisory control and data acquisition (SCADA), distributed control systems (DCS), and programmable logic controller (PLC) systems to aid in monitoring and control of the plant, business computer networks to manage its financial systems, and asset management systems to operate effectively. Because of increasing reporting demands by regulatory oversight agencies and tightening of plant budgets, with resultant staff reductions at most facilities, few can operate for long without fully functioning information systems.

2.0 UTILITY CYBER NETWORKS

2.1 Business Networks

Utility business networks are similar to other business networks found in corporate offices, government offices, and other commercial enterprises. They rely on commercial hardware, software, and wide area networks (WAN) for communications with the outside world and local area networks (LAN) for internal communications. As stated in the introductory paragraph, business networks include all software applications and databases that are essential to the utility's enterprise, including

- Financial programs for accounting, payroll, and human resource management;
- Laboratory information management systems;
- Geographic information systems;
- Enterprise asset management systems; and
- Intranet/Internet to allow customers and employees to interact around the clock with the utility.

2.2 Control Networks

Plant control systems share the same underlying operating systems that are used in the business network. Good design practice separates control networks from business networks by using separate network cables and placing appropriate firewalls and demilitarized zones between the two systems (see Figure 7.1).

Current technology advances with open systems and increasing demand for information are driving tighter connectivity between the two networks. An increasing number of utilities are now leveraging the wealth of process data available from process controllers to provide feedback to the business system. For example, a utility can link their process equipment run times to an asset management system to schedule maintenance task on various equipment. Typical control network equipment may include

- Computers (SCADA servers, operator workstations, and programming workstations);

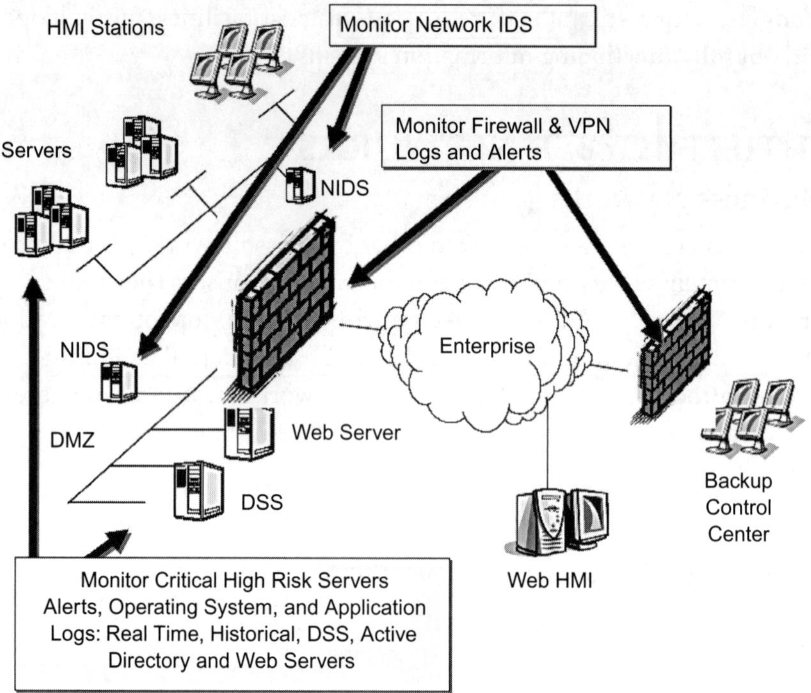

FIGURE 7.1 Cyber-security monitoring (ISA, 2004c; Copyright © 2004 ISA. Reprinted by permission. All rights reserved).

- Networking equipment such as routers, switches, hubs, firewalls, modems, and serial interfaces;
- Communication mediums such as network cables, telephone lines, wireless hardware, and antennas; and
- Control equipment such as PLCs, remote terminal units (RTUs), and DCS.

2.2.1 *Proprietary Platforms*

Many utilities have control systems that rely on vendor-developed systems. These systems were developed by manufacturers to provide plant process control systems and SCADA systems to their customers using available technology and communications. For example, in the 1960s, early SCADA systems used RTUs with custom circuitry and proprietary communication protocols to communicate with a remotely located (typically a central monitoring location) master terminal unit (MTU). Operator interfaces for these early SCADA systems were based on cathode-ray-tube displays with custom human machine interface (HMI) software written in assembly languages (Shaw, 2006). Communication protocols, in many cases, required leased long-distance telephone lines in combination with privately owned radio systems (Boyer, 2004). These proprietary RTU systems dominated the U.S. market until the late 1980s, at which time the first "micro" PLCs were introduced. The nature of these proprietary systems made them less vulnerable to malicious attacks as the information on how to access and modify them required knowledge and expertise of the specific communication protocols for each system and was not widely disseminated or easily obtainable.

2.2.2 *Nonproprietary Platforms*

Today's SCADA and process control systems for water and wastewater use "open architecture," which relies on microprocessor technology for the control hardware and standard Intel (Intel Corporation, Santa Clara, California) or other manufacturer personal computers for the operator interface with Microsoft (Microsoft Corporation, Redmond, Washington) Windows-based HMI software. The most recent trend is to provide remote monitoring capability via Web-based technology. In this case, a browser such as Microsoft Internet Explorer displays hyper text markup language (HTML) pages from a Web server that dynamically creates the Web page using real-time data collected by SCADA. These pages are then published on the LAN of the water or wastewater system operator or, if desired, on the Internet.

Although Microsoft Windows-based systems are prevalent in water and wastewater facilities, other operating systems are also used in some systems (i.e., Linux and UNIX). Because of their cost advantage, PLCs are now commonly used as RTUs and in process control applications in water and wastewater plants. However, unlike RTUs, PLCs are able to perform control of remote sites without the direction of a master. Programming of these PLCs is done using international standards such as International Electrotechnical Commission (IEC) (Geneva, Switzerland) 61131–3 (IEC, 2003). This standard is the international standard for programmable controller programming languages. As such, it specifies the syntax, semantics, and display for the following suite of PLC programming languages:

- Ladder diagram,
- Sequential function charts,

- Function block diagram,
- Structured text, and
- Instruction list.

Communication links have also been standardized and rely on one of several network protocols. The most common communication protocol for business networks and process control system networks (PCS) is Ethernet transmission control protocol/ Internet protocol. Although a number of standard protocols are used at the control level of PCS including ControlNet, DeviceNet, Ethernet/Internet protocol, Modbus RTU, Modbus/transmission control protocol, Fieldbus, and others. The prevalence of these nonproprietary protocols in business systems has increased the risk of cyber attacks to such systems because the information is in the public domain and readily available.

3.0 CYBER SECURITY THREATS

3.1 Policy

To protect the aforementioned plant computer systems from threats, utilities must develop policies. The technical knowledge, skills, and tools required to penetrate IT and plant systems are widely available. It is vital that security of plant computer systems be approached as a collaborative effort between IT and plant engineering, maintenance, and operations personnel. As a starting point, the creation of a cyber security plan in context with a physical security plan must be developed. Such a plan must provide the policies, procedures, and direction for changes to the system and access issues to minimize intrusion risk and insider malfeasance. The plan should include

- Access to all information and control systems (both physical and network access);
- Password policies;
- Limits on information flow between business and control networks;
- System documentation;
- Strict control and elimination of unauthorized wireless or modem connections;
- Disaster recovery plan;

- Incident response; and
- Security improvements.

3.2 Procedures

Once a plan is in place, specific procedures must be developed to implement the plan. The types of procedures developed must address all policy issues in a standardized documented format compatible with the utility's documentation policies. Various guides are available to help develop these documents, including government, standards organization, and private industry. In *Cyber Security for SCADA Systems*, Shaw (2006) developed an actual list of procedures that can be used as a guide.

3.3 Training

To effectively implement security policies and ensure that all personnel are on board with the plan, a cyber security training program must be in place. Training sessions help review security procedures and remind users of their role in carrying out the designated policies. Topics for training sessions should include

- Password protection—users should be instructed not to share passwords with others as intruders have been known to pose as administrators to obtain passwords from unsuspecting users. Passwords should also not be written down. Remote-connected machines should have password protection;
- Log-out procedures for operators when leaving control rooms or other locations where HMI workstations are located;
- Regular analysis of log files by network administrators to detect unauthorized activities; and
- Authorization of wireless and wired connections to control networks or between control and business networks.

3.4 Configuration Management

Process control systems require changes during their lifetimes to meet a utility's operational needs. It is imperative that the procedures developed for cyber security include specific rules for making such changes. For example, HMI software updates must be carried out periodically when new releases are available from the manufacturer to increase functionality or address known deficiencies. Programming changes also need to be made by authorized plant personnel to address operational changes. These changes must be properly documented and implemented and must include program-version control and backup methods so that the plant can revert back to the unchanged version in case problems develop with the new program.

4.0 OPERATIONS

4.1 Intrusion Defense

A number of cyber security intrusion defenses are available from many sources. Software tools are available (Shaw, 2006; Teumin, 2005) as intrusion-detection packages that "watch" resources and programs looking for anomalous program behavior. The tools are otherwise known as *host-based intrusion detection systems*. A variation of this is network-based intrusion-detection systems technology, which consists of one or more computers placed on the LAN/WAN.

4.2 Internet Intrusion

Enterprise Internet security is typically provided by a utility's IT department. The IT department, in some cases, may need specialized training to maintain security at the network's Internet gateway. If a utility's staff does not have this specialized training, then a consultant should be brought in to provide the service.

4.3 Telephone System Intrusion

As stated earlier, many process control systems' SCADA servers and some business systems' remote-access servers still use modems for communications. Telephone systems are vulnerable to unauthorized access through such modem connections.

4.4 Wireless Intrusion

Many utilities rely on wireless transmission for communications between remote SCADA components and the central monitoring system for monitoring and control. These data interchanges are typically encrypted broadcasts that can intercept potentially harmful information and retransmit it with altered information.

4.5 Intrusion Detection

In developing a cyber security plan, the International Society of Automation (ISA) (Research Triangle Park, North Carolina) Technical Report TR99.00.02, article 18.11, recommends the following: "The plan should provide for intrusion detection at an appropriate level for the system, which can range from detecting hardware and physical intrusions to detecting unauthorized remote access or activities. Intrusion detection may incorporate process models, cross correlation between redundant or diverse data, and other techniques to assess the validity of data. The intrusion detection system should also provide appropriate notification and/or response to intrusion detection" (ISA, 2004a).

5.0 DESIGN

5.1 Internet Intrusion Protection Design Practice

Numerous techniques and devices are available for providing Internet intrusion protection, including a virtual private network to prevent unauthorized access into the network from the Internet and firewall "stateful" packet inspection or "proxy" servers. Stateful packet inspection, also referred to as *dynamic packet filtering*, is a firewall architecture that works at the network layer. Unlike static packet filtering, which examines a packet based on the information in its header, stateful inspection tracks each connection traversing all interfaces of the firewall and makes sure they are valid. A stateful firewall may examine not just the header information, but also the contents of the packet up through the application layer to determine more about the packet than just information about its source and destination. A stateful inspection firewall also monitors the state of the connection and compiles the information in a state table. Because of this, filtering decisions are not only based on administrator-defined rules (as in static packet filtering), but also on context that has been established by prior packets that have passed through the firewall. As an added security measure against port scanning, stateful inspection firewalls close off ports until connection to the specific port is requested (Webopedia.com, 2009a).

Proxy server refers to a server that sits between a client application such as a Web browser and a real server. It intercepts all requests to the real server to see if it can fulfill the requests itself. If not, it forwards the request to the real server.

Although they both relate to network security, an intrusion detection system (IDS) differs from a firewall in that a firewall looks out for intrusions to stop them from happening. The firewall limits access between networks to prevent intrusion and does not signal an attack from inside the network. An IDS evaluates a suspected intrusion once it has taken place and signals an alarm. An

IDS also watches for attacks that originate from within a system (see Figure 7.1) (Webopedia.com, 2009b). However, users should be aware of the limitations of IDS, which include

- *Noise*—noise can severely limit an IDS's effectiveness. Bad packets generated from software bugs, corrupt domain name system data, and local packets that escaped can create a significantly high false-alarm rate.
- *Too few attacks*—it is not uncommon for the number of real attacks to be far below the false-alarm rate. Real attacks are often so far below the false-alarm rate that they are often missed and ignored.
- *Signature updates*—many attacks are geared for specific versions of software that are typically outdated. A constantly changing library of signatures is needed to mitigate threats; outdated signature databases can leave an IDS vulnerable to new strategies.

For an overview of IDS and their capabilities, the white paper, *An Introduction to Intrusion Detection Assessment for System and Network Security Management,* is available at http://www.icsa.net/services/consortia/intrusion/intrusion.pdf. For a survey of commercially available IDS that allow one to easily compare features, Los Alamos National Laboratory (Los Alamos, New Mexico) has published *Intrusion Detection System Product Survey* (Jackson, 1999). Additional information on IDS can also be found in *Guide to Intrusion Detection and Preventive Systems (IDPS)* (NIST, 2007). It is recommended that security specialist services be used to periodically evaluate IDS and firewall effectiveness.

5.2 Telephone Intrusion Protection Design Practice

Recommendations to protect against telephone intrusions include

- Policies to prevent or control remote-access modems;
- Use of commercial telephone-scanning software to identify unauthorized access;
- When using authorized modem connections, consideration of encryption commands to prevent interference from attackers; and
- Equipping all SCADA modems with "lock-and-key" hardware devices, with keys provided only to trusted personnel and vendors.

5.3 Wireless Intrusion Protection Design Practice

To mitigate risks associated with wireless intrusions, utilities should consider the following factors:

- Use "hardened" lockable enclosures for all remote control system units as many of these units are located in remote areas subject to vandalism.
- Provide signal supervision and tamper alarms to detect loss of signals and unauthorized entries.
- Use encryption for radio traffic between RTUs and master stations.
- Use frequency hopping spread spectrum radios when geographical area topology permits.
- Turn off "beaconing" and minimize reception area through a combination of antenna type and wireless-access point configuration. Beacons are small packets whose continuous transmission advertises the presence of a base station (access point) beacon. The mobile units sense the beacons and attempt to establish a wireless connection.

6.0 FEDERAL GOVERNMENT INITIATIVES FOR PROTECTION OF INFORMATION TECHNOLOGY NETWORKS

6.1 Department of Homeland Security

The Homeland Security Act of 2002 is a federal policy established by the U.S. Department of Homeland Security (DHS) as the focal point for coordinating activities to protect computer systems that support our nation's critical infrastructures.

In June 2003, DHS established the National Cyber Security Division (NCSD), under its Information Analysis and Infrastructure Protection Directorate, to serve as a national focal point for addressing cyber security issues and to coordinate implementation of the cyber security strategy. National Cyber Security Division also serves as the government lead on a public/private partnership supporting the U.S. Computer Emergency Response Team (US-CERT) and as the lead for federal government incident response.

The Department of Homeland Security and Idaho National Labs have recently completed an agreement for the ISA Automation Standards Compliance Institute to distribute the Control Systems Cyber Security Self-Assessment Tool (CS^2SAT). This self-assessment tool is an excellent first step toward a larger security program. The CS^{2}SAT provides users with a systematic and repeatable approach for assessing the cyber-security posture of their industrial control system networks. The CS^2 SAT is a robust software application that was designed with input from DHS's National Cyber Security Division through a joint research effort between the Water Environment Research Foundation (Alexandria, Virginia), American Water Works Association Research Foundation (Denver, Colorado), The National Institute of Standards and Technology, and other experts from laboratories across the United States. The CS^{2}SAT provides users with a systematic and repeatable approach for assessing the cyber-security posture of their industrial control system networks. This desktop software tool guides users through a step-by-step process to assess their control system network security practices against recognized industry standards. The output from the CS^{2}SAT is a prioritized list of recommendations for improving the cyber-security posture of the organization's industrial control systems environment. The CS^{2}SAT derives the recommendations from a database of cyber-security standards and practices that have been adapted specifically for application to the industrial control system architecture and components. Each recommendation is linked to a set of actions that can be applied to enhance cyber-security controls (visit www.us-cert.gov for a US-CERT CS^{2}SAT fact sheet).

6.2 National Institute of Standards and Technology

The National Institute of Standards and Technology (NIST) (Gaithersburg, Maryland) is working with process control end users, vendors, and integrators to improve the IT security of network digital control systems used in industrial applications. To address security requirements for industrial process control systems and components, NIST formed the Process Control Security Requirements Forum (PCSRF) (NIST, 2006) in the spring of 2001. The NIST-led PCSRF is a working group of users, vendors, and integrators in the process control industry that is addressing cyber-security requirements for industrial process control systems and components, including SCADA, DCS, PLCs, RTUs, and intelligent electronic devices.

6.3 U.S. General Accounting Office

The U.S. General Accounting Office (GAO) is known as the investigative arm of Congress and is a congressional watchdog. The GAO supports Congress in meeting its constitutional responsibilities and helps improve the performance and accountability of the federal government for the benefit of the American people.

In May 2005, GAO published a report entitled, *Critical Infrastructure Protection, Department of Homeland Security Faces Challenges in Fulfilling Cyber Security Responsibilities*. From this GAO report, many cybersecurity technologies that can be used to protect critical infrastructures from cyber attack are currently available. These technologies can help to protect information that is being processed, stored, and transmitted in the networked computer systems that are prevalent in critical infrastructures. An overall cybersecurity framework can assist in the selection of technologies for CIP (Critical Infrastructure Protection). Such a framework can include (1) determining the business requirements for security; (2) performing risk assessments; (3) establishing a security policy; (4) implementing a cybersecurity solution that includes people, processes, and technologies to mitigate identified security risks; and (5) continuously monitoring and managing security. Ultimately, the responsibility for protecting critical infrastructures falls on the critical infrastructure owners (GAO, 2005).

Another GAO report entitled *Critical Infrastructure Protection, Challenges in Securing Control Systems*, found that control systems can be vulnerable to a variety of attacks, examples of which have already occurred (GAO, 2003). Successful attacks on control systems could have devastating consequences, such as endangering public health and safety, damaging the environment, or causing a loss of production, generation, or distribution of public utilities. Securing control systems poses significant challenges, including technical limitations, perceived lack of economic justification, and conflicting organizational priorities. However, several steps can be taken now and in the future to promote better security in control systems, such as implementing effective security management programs and researching and developing new technologies. The government and private industry have initiated several efforts intended to improve the security of control systems (See Figure 7.2 for an illustration of a control system).

6.4 U.S. Environmental Protection Agency

The U.S. Environmental Protection Agency (U.S. EPA) has been an active participant in addressing cyber security issues in the water industry and has produced a number of reports on the topic. The most current report as of the writing of this publication is the *2008 Annual Update to the Water Sector-Specific Plan* (U.S. EPA, 2008). As quoted in its introduction, "this report provides an update to the Water Critical Infrastructure and Key Resources (CIKR) Sector-Specific Plan (SSP), as input to the National Infrastructure Protection Plan (NIPP). This update, based on guidance

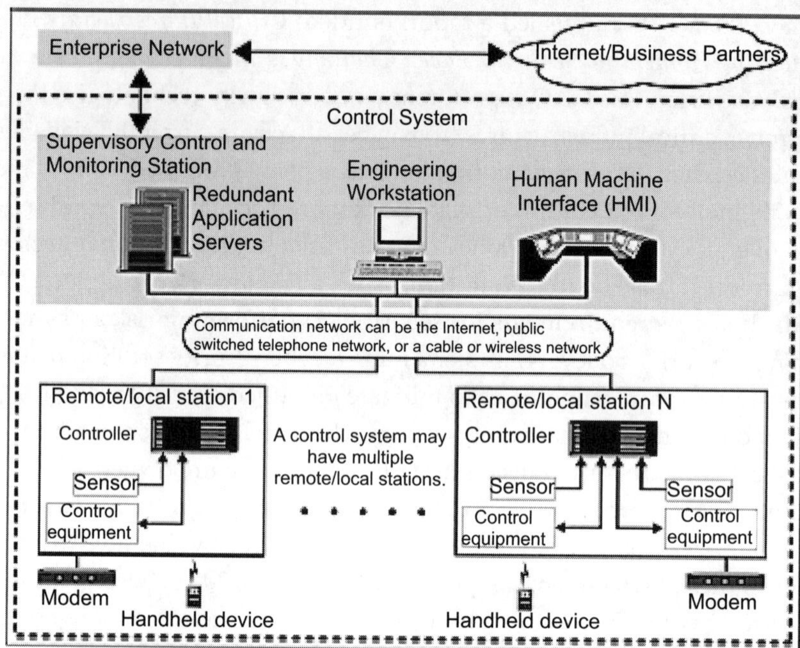

FIGURE 7.2 Control system (U.S. General Accounting Office, 2003).

issued by the U.S. Department of Homeland Security (DHS), provides a summary of advances in the processes set out in the SSP and the achievement of milestones. It is intended to inform Water Sector partners, stakeholders, and Federal, State, local governments, tribes, and other interested parties about updates to key CIKR protection efforts in the Water Sector. The U.S. Environmental Protection Agency (EPA) prepared this update in collaboration with the Water Sector Coordinating Council (WSCC) and the Government Coordinating Council (GCC), two critical vehicles that were established under the NIPP to provide a focal point of interaction respectively with the water industry and other Federal and State partners."

Numerous tools are available to water- and wastewater-sector utilities to help them address water- and wastewater-sector security issues, including cyber security. One such free tool is the *Emergency Response Tabletop CD-ROM Exercises for Drinking Water and Wastewater Systems* (U.S. EPA, 2005). This tool also helps utilities in preparing emergency response procedures (ERPs). Per the report, "the CD-based tool contains tabletop exercises to help train water and wastewater utility workers in preparing and carrying out ERPs. The exercises provided on the CD can help strengthen relationships between a water supplier and their emergency response team (e.g., health officials; laboratories; fire; police; emergency medical services; and local, State, and Federal officials). Users can also adapt the materials for their own needs. Users can choose from five basic event types: intentional contamination, security breach, cyber security, physical attack, and interdependency. The exercises also allow water suppliers to test their ERPs before an actual incident occurs" (U.S. EPA, 2005).

7.0 NONGOVERNMENT INITIATIVES FOR PROTECTION OF INFORMATION TECHNOLOGY NETWORKS

7.1 Water Environment Federation

In conjunction with the American Water Works Association (AWWA) (Denver, Colorado) and the American Society of Civil Engineers (Reston, Virginia), WEF has produced a guideline document entitled *Interim Voluntary Security Guidance for Wastewater/Stormwater Utilities* (WEF, 2004).

7.2 American Water Works Association

The Water Sector Coordinating Council Cyber Security Working Group has prepared a comprehensive document to address cyber security in the water sector. The document, entitled *The Roadmap to Secure Control Systems in the Water Sector*, has been developed to help water utilities develop cyber-security strategies and is sponsored by DHS and AWWA. The document's vision statement summarizes this objective as follows: "In 10 years, Industrial Control Systems for critical applications will be designed, installed, and maintained to operate with no loss of critical function during and after a cyber event" (WSCC CSWG, 2008).

7.3 International Society of Automation

In 2004, under its ISP 99 standards development efforts, ISA published two guideline documents for helping utilities securing their network. The first, *Security Technologies for Manufacturing and Control Systems*, stated in its introduction, "this ISA technical report provides an evaluation and assessment of many current types of electronic security technologies and tools that apply to the Manufacturing and Control Systems environment, including development, implementation, operations, maintenance, engineering, and other user services. It provides guidance to manufacturers, vendors, and security practitioners at end-user companies on the technological options for security these systems against electronic (cyber) attack..." (ISA, 2004b).

A follow-up report entitled *Integrating Electronic Security into the Manufacturing and Control System Environment* (ISA, 2004a) was subsequently released. As stated in its introduction, "this document...provides guidance to Manufacturing and Control System users (including operations, maintenance, engineering, and other user services)...on how to provide adequate electronic (cyber) security for these systems. It focuses on the planning, developing and implementing activities involved with a comprehensive program for integrating security into the Manufacturing and Control Systems environment. The program includes requirement, policies, procedures, and practices increase ranging from risk analysis to management of change and compliance auditing" (ISA, 2004a).

7.4 British Columbia Institute of Technology

In 2004, the British Columbia Institute of Technology (BCIT) (Burnaby, British Columbia, Canada) released one of the first ever global reports on industrial cyber security titled, *The Myths and Facts behind Cyber Security Risks for Industrial Control Systems* (Byres and Lowe, 2004). This report reveals a tenfold increase in successful cyber attacks on process and SCADA systems since 2000. Many of the attacked systems were responsible for the operation of critical services such as electricity, petroleum production, nuclear power, water, transportation, and communications. The study also highlights the significant safety, environmental, reputational, and financial risks that organizations are running every day by failing to adequately address the threat of cyber attacks on their plants and factories. Of those organizations that put a figure on the effect of cyber attacks on their process control and automation systems, 50% experienced financial losses of more than $1 million. The report was produced jointly by security experts at BCIT and PA Consulting Group (London, U.K.).

8.0 REFERENCES

Boyer, S. A. (2004) *SCADA Supervisory Control and Data Acquisition*, 3rd ed.; ISA: Research Triangle Park, North Carolina; pp 22–33.

Byres, E.; Lowe, J. (2004) *The Myths and Facts Behind Cyber Security Risks for Industrial Control Systems;* British Columbia Institute of Technology: Burnaby, British Columbia, Canada, and PA Consulting Group: London, U.K.

International Electrotechnical Commission (2003) *Programmable Controllers, Part 3 Programming Languages, IEC-61131–3,* 2nd ed.; International Electrotechnical Commission: Geneva, Switzerland.

International Society of Automation (2004a) *Integrating Electronic Security into the Manufacturing and Control System Environment;* ISA-TR99.00.02–2004; International Society of Automation: Research Triangle Park, North Carolina.

International Society of Automation (2004b) *Security Technologies for Manufacturing and Control Systems;* ISA-TR99.00.01–2004; International Society of Automation: Research Triangle Park, North Carolina.

International Society of Automation (2004c) Intrusion Detection and Cyber Security Monitoring of SCADA and DCS Networks. Paper presented at ISA Automation West; Long Beach, California, Apr 27–28; International Society of Automation: Research Triangle Park, North Carolina.

Jackson, K. (1999) *Intrusion Detection System Product Survey.* http://lib-www.lanl.gov/la-pubs/00416750.pdf (accessed Aug 2009).

National Institute of Standards and Technology (2006) ISD Process Control Security Requirements Forum (PCSRF) Web Page. http://www.isd.mel.nist.gov/projects/processcontrol (accessed Aug 2009).

National Institute of Standards and Technology (2007) *Guide to Intrusion Detection and Preventive Systems (IDPS),* Special Publication 800–94. csrc.nist.gov/publications/nistpubs/800–94/SP800–94.pdf (accessed Aug 2009).

Shaw, W. T. (2006) *Cyber Security for SCADA Systems;* PennWell Corporation: Tulsa, Oklahoma.

Teumin, D. J. (2005) *Industrial Network Security;* ISA: Research Triangle Park, North Carolina; pp 85–87.

U.S. Environmental Protection Agency (2005) *Emergency Response Tabletop CD-ROM Exercises for Drinking Water and Wastewater Systems;* EPA-817-C-05–001. http://cfpub.epa.gov/safewater/watersecurity/trainingcd.cfm (accessed Oct 2009).

U.S. Environmental Protection Agency (2008) *2008 Annual Update to the Water Sector—Specific Plan;* EPA 817-K-08–002. http://nepis.epa.gov/EPA/html/DLwait.htm?url=/Adobe/PDF/P1002IC7.PDF (accessed Oct 2009).

U.S. General Accounting Office (2003) *Critical Infrastructure Protection, Challenges in Securing Control System;* GAO-04–140T. www.gao.gov/cgi-bin/getrpt?GAO-04–140T (accessed Aug 2009).

U.S. General Accounting Office (2005) *Critical Infrastructure Protection, Department of Homeland Security Faces Challenges in Fulfilling Cyber security Responsibilities;* GAO-05–434. http://www.gao.gov/new.items/d05434.pdf (accessed Aug 2009).

Water Environment Federation (2004) *Interim Voluntary Security Guidance for Wastewater/Stormwater Utilities;* Water Environment Federation: Alexandria, Virginia.

Water Sector Coordinating Council Cyber Security Working Group (2008) *Roadmap to Secure Control Systems in the Water Sector.* http://www.awwa.org/files/GovtPublicAffairs/PDF/WaterSecurityRoadmap031908.pdf *(accessed Aug 2009).*

Webopedia.com (2009a) Stateful Inspection. http://www.webopedia.com/TERM/S/stateful_inspection.html (accessed Oct 2009).

Webopedia.com (2009b) Virtual Private Network (VPN). http://www.webopedia.com/TERM/V/VPN.html (accessed Oct 2009).

9.0 SUGGESTED READINGS

U.S. Department of Homeland Security, U.S. Computer Emergency Readiness Team, *Control System Security Program* Home Page. http://www.us-cert.gov/control_systems/ (accessed Aug 2009).

U.S. Department of Homeland Security, U.S. Computer Emergency Readiness Team, *Control System Cyber Security Self Assessment Tool.* http://csrp.inl.gov/Self-Assessment_Tool.html (accessed Aug 2009).

U.S. Department of Human Services (2006) National Infrastructure Protection Plan. http://www.dhs.gov/xprevprot/programs/editorial_0827.shtm (accessed Dec 2009).

U.S. Environmental Protection Agency (2007) *Water—Critical Infrastructure and Key Resources Sector-Specific Plan as Input to the National Infrastructure Protection Plan,* 817 –R-07–001. http://www.epa.gov/safewater/watersecurity/pubs/plan_security_watersectorspecificplan.pdf (accessed Oct 2009).

Chapter 8

Organizational Aspects of Information Technology

(continued)

1.0 HOW WELL IS TODAY'S UTILITY BEING SERVED BY INFORMATION TECHNOLOGY?

The water and wastewater department of a major Canadian municipality recently rated its partnership with corporate information technology (IT) at 3.1 against a target score of 5.0, with possible answers ranging from 1 (strongly disagree) to 6 (strongly agree). Only one corporate services provider in the city was ranked lower in the survey. According to many respondents, the perceived limitations of IT existed despite internal IT charges that were considered to be high.

Upon further analysis, IT's poor rating was because of (1) the lack of clear assignment of responsibility for IT decision making, business applications, and data management; (2) the lack of effectiveness of corporate IT services in meeting water and wastewater customer needs; and (3) the lack of recognition of IT as an important strategic contributor to the success of water and wastewater services. As a result, the city's IT function was characterized by ad hoc decentralization of IT resources, unmanaged growth of standalone business applications and databases, and a corporate IT organization with a limited focus on its customers.

1.1 Unclear Assignment of Responsibility for Information Technology Decisions and Functions

Responsibility for decision making in the areas of IT principles and policies, IT architecture, IT infrastructure, business applications, and IT investment was not clearly assigned in the aforementioned Canadian municipality. This made it difficult for the water and wastewater department to gain the full potential offered by IT solutions. In the business application area, for example, the maintenance and use of applications and related data were sometimes performed outside the water and wastewater department and sometimes outside corporate IT. Examples of shared responsibility for application included computerized maintenance management systems (CMMS), which were the responsibility of both the public works and water and wastewater departments; geographic information systems (GIS) and hydraulic modeling software, which were the responsibility of the planning department; and project management software, which was the responsibility of the engineering department. In each case, joint, indirect, or sometimes unclear responsibility resulted in a lack of priority being placed on making water- and wastewater-specific updates and improvements on a timely basis. Corporate IT would have improved its service by developing clear application and data architectures and providing stewardship of a robust IT governance structure that documents standard ways to make decisions and allocates responsibilities between departments and corporate IT.

1.2 Lack of Customer Focus by Corporate Information Technology

Further analysis behind the survey at the Canadian municipality showed that according to the water and wastewater department, the culture in corporate IT had evolved away from the specific local service needs of its customers and turned to the corporate desktop, major corporate applications, and a focus on technology. It was suggested that corporate IT's pending application of a service-oriented architecture (SOA) approach was not matched by either a service-oriented attitude or a customer-facing organizational structure. Service-oriented architecture is a business-centric IT architectural approach that supports integrating a business as linked, repeatable business tasks, or services. Using the SOA approach can help municipalities and utilities find value at every stage of the SOA continuum, from departmental projects to enterprise-wide initiatives. The full value of SOA, however, can only be achieved in municipalities or utilities where the IT organization is highly focused on serving its customers.

Evidence of the corporate IT shift away from direct involvement with its departmental customers could be found in the way corporate IT resources were organized. Three groups of centralized IT staff were clearly visible on the organization chart, controlling the corporate desktop, prioritizing and implementing requests for upgrades to major corporate applications, and keeping corporate networks and hardware running. Corporate IT would have improved its service by providing customer-focused resources such as customer-relationship managers in addition to the more traditional help-desk support team and IT project managers.

1.3 Ad Hoc Decentralization of Information Technology Resources

The perceived inability of corporate IT at the Canadian municipality to provide the right resources and effectively respond to its customers' needs caused many departments, including water and wastewater, to hire their own IT resources. Unfortunately, this hiring was done in the absence of a municipal-wide IT strategic plan, without the guidance of an enterprise architecture, and outside the auspices of an appropriate IT governance structure. At the time the survey was conducted at the Canadian municipality, a substantial number of IT resources were present throughout front-line departments like water and wastewater. In a number of cases, there was an overlap in responsibilities between corporate IT and distributed IT resources and, in other cases, gaps existed. This resulted in inefficient and ineffective allocation of resources and caused confusion and consternation between the departments and corporate IT. Corporate IT would have improved its service by providing a clear organizational architecture and IT governance structure.

1.4 Other Opportunities for Information Technology to Better Serve Water and Wastewater Utilities

There are a number of other opportunities for IT to better serve municipal water and wastewater departments and utilities. In some municipalities, the battle for responsibility for IT systems and applications is in process control and supervisory control and data acquisition (SCADA) systems, CMMS, GIS, project management systems, telecommunications, and, more recently, asset management systems. In large utilities in Canada, for example, process control systems are typically the responsibility of operations and maintenance, whereas financial and human resources systems are typically under the auspices of corporate IT. In the United States, it is estimated that process control systems are the responsibility of IT in approximately 50% of large utilities.

Regardless of who is responsible for what system, the most important factor in their success is that responsibilities are clearly understood and that relationships between various stakeholders are clearly defined.

In some organizations, there is insufficient strategic coordination of applications, even within water and wastewater utilities. Several large utilities in Canada, for example, use two different CMMS applications, one to maintain treatment plant assets for water and wastewater treatment and the other to manage distributed network assets in the network operations. From an individual business-unit perspective, this may have produced the right results; however, from an IT and business-wide perspective, the solution was sub-optimal. Corporate IT would have improved its service by providing the strategic direction and architecture to guide the utility to the optimal overall solution.

Regardless of the location of IT resources or the allocation of responsibilities for IT services, opportunities to improve the conversation between engineers, operators, IT professionals, and financial specialists remain in many utilities. Information technology strategic plans still do not always reflect the business requirements as effectively as desired, utility executive management teams still do not include a chief information officer (CIO), and there are still utility IT organizations with opportunities to organize themselves to provide better focus on the business of their front-line utility customers.

In summary, there are typically still plenty of opportunities for IT organizations to better serve today's utility. The best way for IT organizations to realize those opportunities is to meet the organizational and governance challenges head-on in a strategic, systematic way.

2.0 ORGANIZATIONAL AND GOVERNANCE CHALLENGES

The organizational and governance challenges a typical utility IT department faces can be significant. A plethora of hardware, software, databases, networks, and other IT systems have to respond to a set of complex business requirements that demand a myriad of resources and skills. Information technology services can range from calibrating instruments in a treatment process to making strategic decisions in the boardroom.

2.1 Top 10 Organizational Challenges Facing Utility Information Technology

The top 10 organizational challenges facing a typical utility IT department today include (1) integrating the utility IT vision with the utility business vision; (2) lack of full appreciation of the strategic significance of IT because of "engineering orientation" of utility management; (3) lack of a strong IT governance framework to guide IT investments and ensure adherence to architectures and policies; (4) lack of an IT "voice" on the utility's executive board; (5) insufficient IT resources that truly understand the utility's operational and business requirements; (6) technologies are changing faster than customer organizations can adopt them; (7) a significant technology awareness and usage gap between the utility's senior management and the latest generation of technology savvy employees; (8) lower IT pay levels in public sector utility organizations than in the rest of the IT industry; (9) attracting and retaining qualified IT resources; and (10) the need for IT staff to work well with varying cultures of employees from planning, engineering, operations, maintenance, and finance.

Other related challenges include IT budget pressure, the politics of off-shoring (especially in the public sector), use of software as a service (SAAS) and other outsourcing, and legislation (e.g., Sarbanes-Oxley). Local factors determine the priority level of each challenge.

In many utilities, the roots of IT challenges can be traced back to more than just organization and governance. Restrictive procurement policies, civil service constraints, and union issues have also been significant hurdles to realizing the full strategic potential of utility IT organizations.

2.2 Municipal Utilities

Information technology challenges are typically greatest for water and wastewater departments governed within larger municipalities that deliver a broad range of public services. Those departments often face tough battles for IT resources, funding, and priorities. In addition, they sometimes need to create and maintain more working partnerships with other front-line service organizations to reflect their own needs in shared applications and data. Examples of this include sharing a CMMS with the roads and highway department, an Enterprise Asset Management System with the finance department, or a GIS with various organizations in public works and corporate services.

In addition, some major municipalities can be organizationally more complex and convoluted. In larger cities in the United States, for example, it is not unusual to find 12 or more levels in a corporate hierarchy, from the city manager to instrument maintenance technician. More levels and more partnerships can mean slower IT response times, less integration, and suboptimal IT business effect. In some municipal water and wastewater departments, the only "in-house" resources dedicated to IT-related technologies are those responsible for instrumentation, control, and automation systems.

A number of leading utilities have adopted a more customer-focused "IT associate" or customer-relationship manager model, where IT resources are co-located with front-line water and wastewater operating groups while reporting into corporate IT departments. However, even this practice is not universal and addresses only one dimension of bridging the gap between IT as a desktop provider to IT as a strategic business partner.

2.3 National Utilities

There are a number of countries where water and wastewater needs are met by a single nationwide utility. These national water and wastewater utilities are often dependent on various government ministries for resources, policy direction and data, and application support. In developing countries, these utilities often have difficulty finding local resources to meet their IT needs. For instance, when these utilities look for solutions by hiring vendors, consultants, and contractors to cover the shortfall, temporary rather than sustainable relief is provided.

At one large national utility in the Caribbean, for example, consultants and contractors have installed a large variety of instrumentation and control equipment during the past 20 years. Unfortunately, much of it was done without considering maintainability and the local capacity to service the equipment. The same can be said for a number of IT applications and technologies at this utility. Utilities in other developing countries often face similar challenges. Indeed, the solution of flying in expensive experts to provide a temporary fix is unlikely to enable IT organizations at those utilities to higher levels of maturity and self-sufficiency.

2.4 Publicly Owned and Privately Operated Utilities in North America

Private companies operating and maintaining utility assets for customer organizations may be dependent on that customer's IT resources for data and access to external IT systems and services. Sometimes, this occurs in an environment where responsibilities for providing and maintaining data, managing databases, and maintaining IT systems are not well documented in operating agreements. They may also have to provide IT resources with the skills to manage a combination of customer-mandated systems, such as CMMS and SCADA, and their own preferred systems. These preferred systems are sometimes used as standard tools by private operators to ensure consistent and efficient service delivery for customers across a region or nation.

2.5 Privately Owned and Operated Utilities

Fully privatized utilities such as those in the United Kingdom have complete responsibility for all aspects of IT required to efficiently and effectively deliver water and wastewater services. Their approach to IT has changed considerably since privatization in 1989, and has also varied greatly between utilities.

During the first post-privatization decade, most of the 10 large private water utilities grew their internal IT organizations and service offerings beyond anything traditional utilities had ever seen. One went as far as developing its own remote terminal unit (RTU) hardware, whereas others created and spun off separate IT organizations that provided services and specialized application software to the originating utility, other U.K. utilities, and utilities in other countries.

This expansion and divestiture of IT caused special challenges for core utility organizations. Should they acquire and use only their own proprietary IT resources and software solutions or could they use other vendors? What IT role would remain in the utility? And how would their service requests be prioritized by a more arms-length IT service provider?

During the second post-privatization decade, a number of large water utilities narrowed the scope of IT products and services they delivered, the business sectors in which they were offered, and the geographic markets in which they serviced. Thames Water, one of the world's largest water and wastewater utilities that serves customers in and around London, England, for example, returned to a focus on delivering core water and wastewater services in its domestic U.K. market, supported by internal services from its IT department. For utilities like Thames Water, organizational challenges in IT are moving back to the more traditional utilities mentioned previously. Others retain their external IT service delivery companies, but focus their efforts on serving the water industry.

2.6 Who Is Responsible for Information Technology at Today's Complex Utility?

A typical utility today has a large number of individuals and groups responsible for IT systems and services that, for purposes of this discussion, include hardware and software used for monitoring and control of water and wastewater processes and equipment. These individuals are skilled in areas such as IT infrastructure management and support, application support, systems management, and possibly project management. Or, they are skilled in planning and engineering, process control and SCADA system management, network optimization, or instrumentation maintenance. Specific roles with responsibility for IT range from instrumentation and control technicians to corporate business architects and CIOs. Figure 8.1 shows the typical distribution of IT responsibilities as they relate to hardware and networks found in today's complex utility.

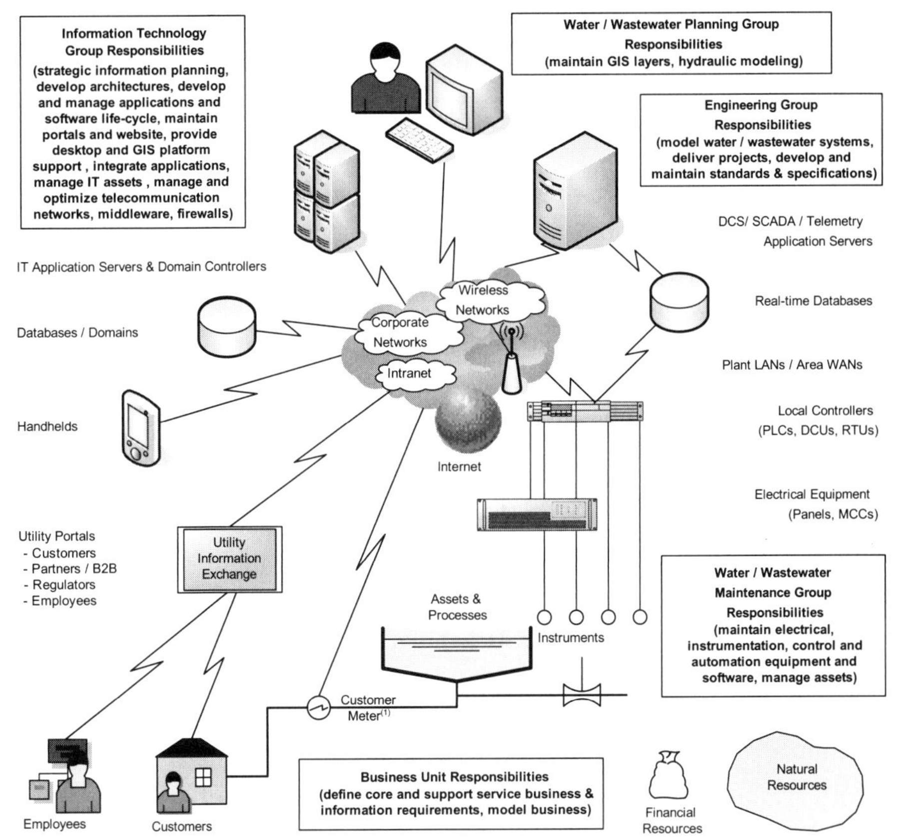

FIGURE 8.1 Responsibility allocation for IT services in today's typical utility.

The well-designed and organized utility IT provider of the future would proactively meet all its challenges and best allocate all IT responsibilities using a strategic and integrated organizational response. Information technology's organizational maturity in terms of culture, management style, customer relationships, communication, partnerships, and organizational structure and IT governance has a major role to play in meeting this response.

3.0 MEETING CHALLENGES

The 8-step process described in this section provides a systematic way for utility IT organizations to undertake a successful transformation from the current "as-is" to the target "to-be" maturity level, organizational structure, and governance framework. It also helps those organizations identify and apply ways to respond to opportunities and challenges. Each of the following steps can be focused and adapted for specific utilities and situations:

(1) Define and characterize the current state;
(2) Identify, discuss, and inculcate the future state;
(3) Develop strategies that leverage workforce and technical trends;
(4) Select your organizational structure;
(5) Fill new positions and build better connections;
(6) Get strategic and raise the profile;
(7) Decide who does what; and
(8) Manage the change.

3.1 Define and Characterize the Current State

An industry-accepted way of defining and categorizing the state of an IT organization is the capability maturity model (CMM). In this manual, the CMM has been adapted to focus on organizational characteristics rather than including all the service-related characteristics. Table 8.1 shows the organizational maturity matrix (OMM), which describes the organizational characteristics of utility IT organizations at various levels of maturity.

To determine how close to fully "mature" an IT organization is, the following questions need to be asked in the context of "How close are we to having a...":

- Collective sense of IT mission, strategic IT future vision, both with a dynamic relationship to the utility's mission and vision?
- Strong culture, tightly defined, aligned with the corporate culture and values, shaped by a commitment to innovation, creativity and excellence, to which change comes naturally?
- Structure that is "soft," networked, customer-oriented, and flexibly integrated with the rest of the utility?
- Leadership team in IT that helps formulate and sponsor new initiatives, regularly reinvent IT, and integrate IT into the utility's strategic direction and lead change?
- Communication with staff and employees that is continuous and informal with periodic formal forums and staff meetings to augment daily communications?

Table 8.1 Organizational maturity matrix for water and wastewater utility IT (adapted from Capability Maturity Matrix).

Organizational maturity	Description	Organizational performance indicators
1 IT tasks	• Reactive work • Task focus • Results dependent on individuals • Least productive	No collective sense of IT mission or IT future vision Culture unidentifiable, widely variable Structure loose, decentralized, decisions based on influence of individuals Management role unclear, inconsistent, may contribute to projects, daily IT tasks Communication with staff and employees inconsistent and infrequent Customer relationships based on connections between individuals, customers seen as passive recipients of IT services, customer experience varies greatly
2 IT projects	• Some routine work • Project focus • Some standard results, but variable between groups • Results delivered by project teams and work groups	Little collective sense of IT mission, typically no strategic IT future vision Culture hard to identify, strong and competing subcultures may exist Structure hierarchical, decisions made based on position-power Management more removed from projects and daily IT functions Communication with IT staff infrequent, often only when something is wrong Customer relationships based on connections between groups and individuals, customers seen as recipients of IT services and participants in IT projects, customer experience varies by project and work group
3 IT business	• Proactive, planned and systematic work • Program and service delivery focus • Consistent, standard results • Results delivered by everyone in IT following methods and standards on projects and other services	Collective sense of IT mission, typically no strategic IT future vision Culture likely strong, well-defined, and shaped by standards and procedures Structure hierarchical and bureaucratic Management remote from daily functions of the organization Communication with IT staff done through a series of formal directives and regular staff meetings or presentations Customer relationships based on standard expectations built over time, customers seen as recipients of IT services, some periodic feedback received, customer experience fairly consistent

(continued)

Table 8.1 (*Continued*).

Organizational maturity	Description	Organizational performance indicators
4 Utility business IT	• Planned work driven by customer needs • Utility focus, IT supports efficient business processes • Monitored and managed performance of IT services • Results delivered by everyone in IT focusing on customer relationships and impact on business performance	Collective sense of IT mission, strategic IT vision, reflects utility's mission, vision Culture strong, tightly defined, shaped by commitment to excellence, some ability to change Structure flat, flexible, customer-oriented, some ability to change Management involved in daily IT functions only as facilitator and coach Communication with IT staff continuous and informal, with periodic directives and staff meetings to augment daily communications Customer relationship based on working partnerships with joint translation of business needs into IT solutions. Feedback is integral part of the relationship, customer experience well managed through surveys, periodic joint sessions Strategic leadership role of IT in utility beginning to take shape through organizational visibility of IT (e.g., direct report to top executive), strategic IT roles (e.g., CIO)
5 Utility strategy IT	• Work driven by innovation, anticipated change, and strategic customer needs • Utility focus, IT introduces new, innovative IT processes and initiatives • Results delivered by everyone in IT focusing on customer relationships, business performance improvement, and jointly inventing the utility of the future	Collective sense of IT mission, strategic IT future vision, dynamic relationship to utility's mission and vision Culture strong, tightly defined, consistent with corporate culture and values, shaped by commitment to innovation, creativity and excellence, change comes naturally "Soft" structure, networked, customer-oriented, flexibly integrated with rest of utility IT leaders help formulate and sponsor new initiatives, reinventing IT, integrating IT into strategic business direction and leading change Communication with staff and employees is continuous and informal with periodic directives and staff meetings to augment daily communications Customer relationship characterized by close working partnerships in all stages of the IT and business solution life cycle. Bidirectional, dynamic feedback is an integral part of the relationship, customer experience continues to improve Strategic leadership role of IT in utility visible, enabled through strategic governance structures that include all utility business leaders, strategic IT roles (e.g., CIO)

- Relationships with customers characterized by close working partnerships at all stages of the IT and business solution life cycle, using multidirectional, dynamic feedback as an integral part of the relationship while each customer's experience continues to improve?
- Visible strategic leadership role for IT in the business of the utility, enabled through strategic governance structures that include all utility business leaders, recognizable by the presence of strategic IT roles (e.g., CIO and vice president of IT)?

Although these questions are all in organizational areas that are difficult to measure, the areas should at the very least be gauged by using a simple, repeatable survey based on the OMM shown in Table 8.1. The process of asking questions and having discussions about survey results provides valuable contributions to positive change. Use of more quantifiable metrics to track outcomes such as productivity increases and improvements in service levels could be applied at utility-wide and team-specific levels. Additional metrics could also be developed as specific opportunities are identified and tracked.

The maturity level of IT organizations in many utilities is low when measured against the model shown in the OMM. It shows organizational performance indicators for each of 5 levels of maturity, including

- Level 1, which focuses on performing IT tasks;
- Level 2, which focuses on delivering IT projects;
- Level 3, which focuses on delivering IT services to deliver on IT's mission and vision;
- Level 4, which focuses on applying IT to deliver water and wastewater services to deliver the utility's mission; and
- Level 5, which focuses on integrating IT as a fundamental strategic element of the utility's future vision.

Many utility IT organizations find themselves at levels 1 or 2, some at level 3, even fewer at level 4, and very few at level 5. Therefore, for most of these organizations, unifying the entire IT response to utility business needs under an umbrella of shared strategic direction and unimpeded teamwork would provide significant benefits. Organizationally, that would require raising the corporate profile of IT, better defining IT roles and responsibilities, and improving the partnerships between corporate

IT resources, IT resources in front-line departments like water and wastewater, and the various functional specialists throughout the utility. It would also require creation of an inextricable link between the IT organization and its services to the delivery of the utility's mission and strategic vision.

3.2 Identify, Discuss, and Inculcate the Future State

The utility of the future would deliver IT through an organization that has adopted all the characteristics of the mature level-5 organization, as described in the OMM and the traditional CMM. It would also anticipate trends in workforce and technology, leverage opportunities provided by the marketplace, and consider strategic implications of global trends on water and wastewater.

The IT organization's vision would take all those elements into account and be fully aligned with the overall utility's vision. Table 8.2 shows the municipal IT vision and mission for the city of Palm County, Florida. The city is focused on the

TABLE 8.2 City of Palm Coast, Florida, IT vision and mission aligned with utility (department) vision and mission.

Water and wastewater utility	Corporate information technology
Vision	**Vision**
As the community grows, so will the utility. The utility department is comprised of a highly trained professional staff that is prepared to provide the best level of service to the community today and into the future. While using the latest technology, along with proven industry standards, we will continue to develop the infrastructure to meet the needs of the growing community.	The information technology and communications department will be a proactive leader, identifying issues and offering innovative solutions to enable city departments to accomplish their goals and provide quality services to our citizens more effectively and efficiently.
Mission	**Mission**
Our mission is to provide safe drinking water and the best wastewater service to our customers at the lowest possible cost while adhering to the strictest guidelines for water quality and environmental protection.	Information technology is committed to serving the business operations of the city by providing enterprise-wide, integrated solutions with emphasis on superior customer service. Ensure effective and efficient use of new and existing technology resources and investments. Exceed internal and external service expectations by implementing leading-edge solutions in line with established *"EGov"* best practices.

contributions IT will make to front-line department-service outputs and program outcomes, innovative contributions to those outcomes, and a commitment to delivering enterprise-wide and leading IT solutions.

For IT departments in nonmunicipal utilities, the vision and mission could be even more specific, referring to residential, industrial, commercial and institutional customers, regulatory reporting, and the use of the utility portal to build on customer relationships through ready access to information.

The principle of involvement is important to follow when creating a vision and mission for the IT organization. If possible, everyone in the IT organization should be involved while keeping the customer in mind. In addition, customers should be involved for input and feedback. Indeed, the more mature utility organization will increase involvement of its customers in strategic planning.

3.3 Develop Strategies That Leverage Workforce and Technical Trends

It is important for utility IT organizations to leverage workforce and technology trends when developing strategies to reach their vision.

3.3.1 Widening Generation Gap

When 115 senior-manager mentors and trainees at a national water and wastewater authority in the Caribbean were recently asked to raise their hands if they had ever used a social networking site, only 1 of more than 40 mentors raised their hand while every trainee raised theirs. Indeed, those under the age of 25 entering the workforce in most countries are using personal, handheld technology every day as a fundamental part of their social life, academic endeavors, and job functions.

The "wired" worker skill set will be a fundamental part of every job applicant for utilities and their IT departments for the next generation. By 2020, many utilities will have replaced baby boomers, representing between 30 and 40% of the current workforce, with the new wired millennial generation. "Baby boomer" is the label given to the generation of workforce born between 1946 and 1962; generation "X" is the generation born between 1963 and 1980, and "millennials" are the generation born after 1981.

3.3.2 Ubiquitous Technology

The increasingly ubiquitous nature of IT in the world and workplace will affect utility IT organizations significantly. New employees will expect to be fully supported by the latest technology and the best decision-making tools. Otherwise, they may choose to work at more technology-savvy organizations. Operators and maintenance workers will expect the availability of process conditions, equipment, and recent laboratory test results on their handhelds as they're doing their rounds. In addition, network technicians will expect to be able to remotely diagnose and even repair any potential problems in corporate and local networks.

Front-line staff will also be able to contribute more of their own skills in configuring technology tools as knowledgeable self-supporting users. This can, in turn, contribute to decentralization of a number of IT services.

3.3.3 *Changing Workforce*

Strategic IT organizations need an understanding of other global trends in the workforce as well including the following: the reduced availability of technical IT staff, both locally and globally; the changing aspirations and attitudes of the IT workforce (Salkowitz, 2008) as a new generation of staff is hired; the changing ethnic mix in the workplace; and the role of outsourcing and "off-shoring" of IT services. These trends will have a significant effect on utility leaders as they decide how to best assign IT responsibilities over the next 20 years.

In the best utilities, IT leaders provide strategic advice on how the business might anticipate and apply advanced technologies, from self-cleaning remotely diagnosed dissolved oxygen probes to embedded state-of-the-asset reporting technologies to keep the business at the forefront of providing value to customers and shareholders.

3.4 Select Your Organizational Structure

There are three recognizable design models for utility IT organizations to consider adopting for their journey to reach the vision. These are

- The *department model*, where most IT services are decentralized and corporate IT is responsible for the desktop, the corporate network, and the applications and IT services not covered by front-line departments. From an application perspective, corporate IT is only responsible for those applications for which distributed departments cannot provide support and have requested assistance. Major applications are the responsibility of front-line departments. For example, financial management systems are the responsibility of the finance department, the operations department is responsible for the operations management system, the laboratory is responsible for the laboratory information management system (LIMS), and operations and maintenance is responsible for CMMS. Most IT financial and human resources are allocated to the line departments rather than corporate IT.

- The *shared model*, where IT services are shared between corporate IT and front-line departments and where corporate IT is responsible for the desktop, the corporate network, and a series of agreed upon enterprise-wide applications and IT services. From an application perspective, corporate IT is responsible for key enterprise-wide applications that are determined to be better managed for the good of the overall utility. Typically, this would include systems such as financial systems, human resource systems, GIS, and enterprise asset management systems. Specialized applications are the responsibility of front-line departments. For example, the operations department is responsible for the operations management system and the laboratory is responsible for the laboratory information management system. Information technology financial and human resources are allocated evenly to the line departments and corporate IT.
- The *enterprise-wide model*, where corporate IT is responsible for creating and managing an enterprise-wide information architecture; setting and enforcing enterprise-wide standards for integration, telecommunication, security, data, and information management; and the provision of all IT services. From an application perspective, corporate IT is responsible for all major applications, including operations management systems, LIMS, and CMMS. From an information perspective, it would take data from SCADA and process control systems, likely handed off in time-based packets, in order for it to be used in all other systems on the integrated network. The remainder of the utility organization would be responsible for defining all their respective departments' business requirements. Most IT financial and human resources are allocated to corporate IT rather than the line departments.

Figure 8.2 shows the aforementioned three organization design options mapped against the OMM. Which option is best is dependent on a number of design factors specific to each utility and the capacity of the IT department and the IT service providers available in the marketplace.

In addition to these three distinct models, there are an infinite number of combinations that can be used. Hybrid models would result from application of the

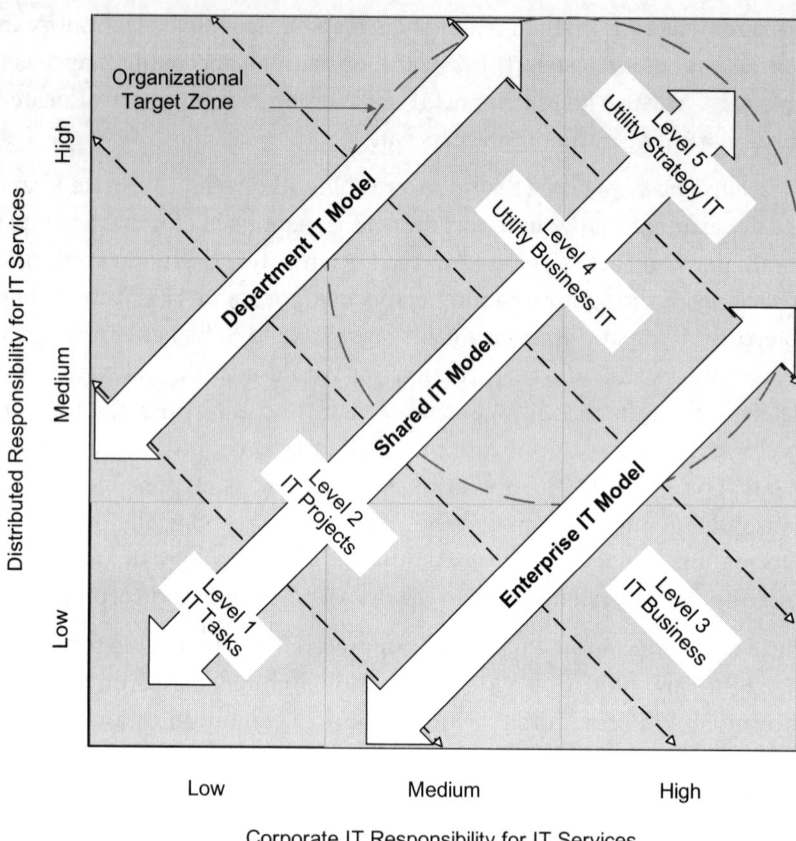

Figure 8.2 Map of organization options for IT in utilities.

more detailed IT decision-making matrix (i.e., establishing a contract/contractor management hub).

As the organization design process moves closer to the preferred high-level structure, decisions also need to be made at the next level of the IT organization. For each natural area of IT expertise, business solution delivery, and customer relationship management, the IT organization would establish teams responsible for centers of expertise, or IT hubs. The types of hubs that could be established include an architecture hub, a hardware environment hub, an e-utility hub, a customer/customer relationship hub, a business solutions hub, a project/program management hub, and a contract/contractor management hub.

The choice of how specific IT responsibilities are allocated across corporate IT, its hubs, and the rest of the utility would be based on a number of design factors. Examples might include

- Internal capacity of IT at the department and corporate IT levels;
- Effect of the selected organization on the ability to
  - Provide for efficient and effective integrated process delivery,
  - Accommodate new projects/programs,
  - Leverage expertise (i.e., the same functions together),
  - Provide a friendly customer interface, and
  - Be responsive to customer requests and needs;
- Manageable reporting structure (e.g., workload balance/reasonable number of reports);
- Level of geographic customer distribution; and
- Ability of the local IT service marketplace to provide resources.

One strategy followed by utility IT organizations to be closer to their customers is the allocation and co-location of IT associates to customer departments. In addition, specific customer-relationship managers can be assigned responsibility for building strong links to front-line departments

3.5 Fill New Positions and Build Better Connections

In the future, there will be a trend in utility IT organizations toward hiring more resources that provide value in the development of business solutions, customer service management, enterprise architecting, and the creation of enterprise-wide architecting frameworks. In addition, these resources would interact and communicate with all internal customers, partners, and external service providers, with clear agreements on expected levels of service.

Competitive salaries, systematic mentorship programs, explicit career path development, and use of the strong marketing allure of the environment and public service will all be required to successfully compete for increasingly scarce IT resources in the future. Meeting the challenge of paying outside traditional pay scales for municipal employees will have to progress beyond current liberal usage of titles. There are too many IT managers in utility organizations whose level of technical expertise and value in the marketplace require payment at a higher salary band. Therefore, the future lies in the creation of municipal pay scales that recognize a technical stream and a management stream of job progression, both of which are calibrated to market conditions.

Figure 8.3 shows roles and interactions that would be characteristic of the mature utility IT organization of the future. It also shows a strategic IT governance council and its potential members, including the CIO.

Decisions on allocating responsibility for service delivery would ultimately drive which IT skills and talent reside inside the organization, where they would reside, and how they would interact with other IT resources. The IT organization

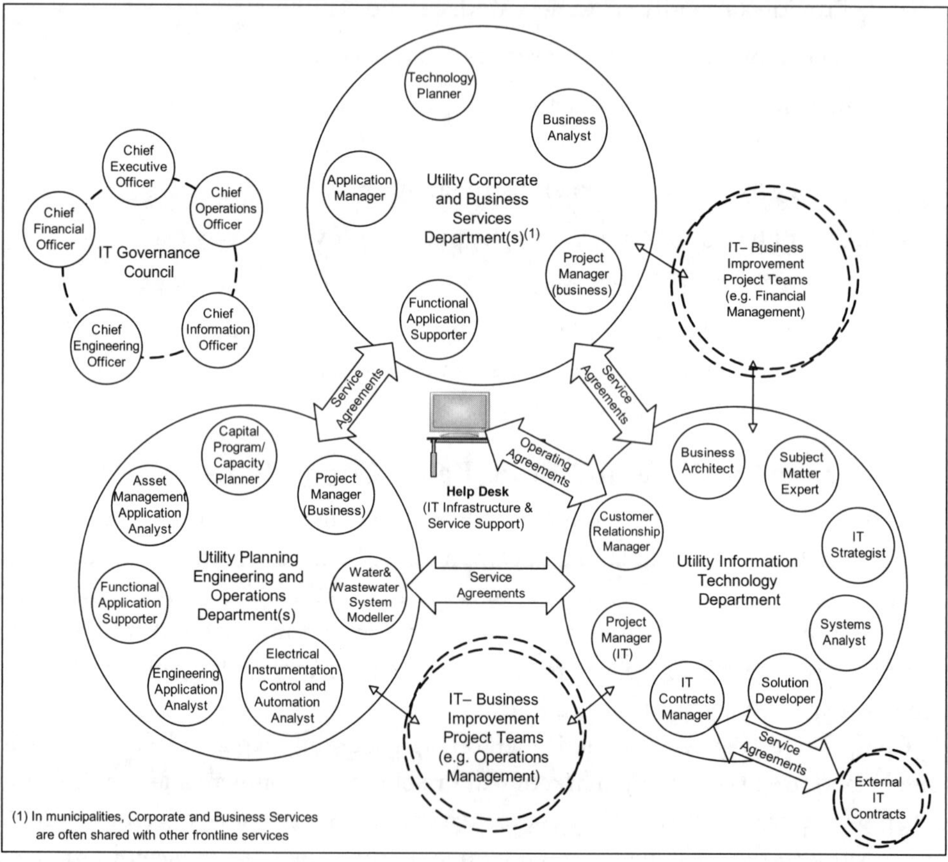

FIGURE 8.3 Roles and relationships map in the mature utility IT organization of the future.

of the future would respond to the utility's business needs and make sure relationships with internal customers and external partners work well. It would then translate those business needs into information, application and technology architectures, integrated solutions, and IT service delivery processes while leading and facilitating the processes to connect with internal customers.

3.6 Get Strategic and Raise the Profile

In the future, IT leaders in the best utilities would provide strategic advice as members of the utility's executive management team. This advice would focus on how IT could contribute to addressing the utility's strategic business challenges, how new technologies could be applied to deliver increasingly valuable services to customers and stakeholders, and how the strategic application of technologies can best leverage applications and other IT resources for the benefit of the entire utility.

There are two ways utility IT organizations could address this challenge in the future. First, they could establish an executive position such as vice president of IT or CIO. Preferably, this position would report directly to the utility chief executive officer, general manager, or president to send a strong message about the importance of IT to the entire utility. In municipalities, it would report to the city manager or chief administrative officer. The IT leader in this position would have a full seat on the utility's executive team to allow ongoing strategic input into the business of the utility.

Secondly, IT leaders would establish an IT governance structure that would complement the corporate governance structure and include an IT forum under executive business sponsorship to govern the development and promulgation of all strategic utility-wide IT frameworks and standards. This forum, sometimes called an *IT governance council*, ensures everyone adheres to the main principle of maximizing IT value for the entire utility enterprise. Figure 8.4 shows a best-practice IT governance design for a municipal water and wastewater utility.

3.7 Decide Who Does What

Information technology leaders of the future would determine who should deliver which IT services in discussion with their executive peers. They would introduce market forces and create healthy competition without undermining the internal IT-service delivery team. This could be done through periodic benchmarking and by maintaining a reasonable balance between internally and externally delivered IT services.

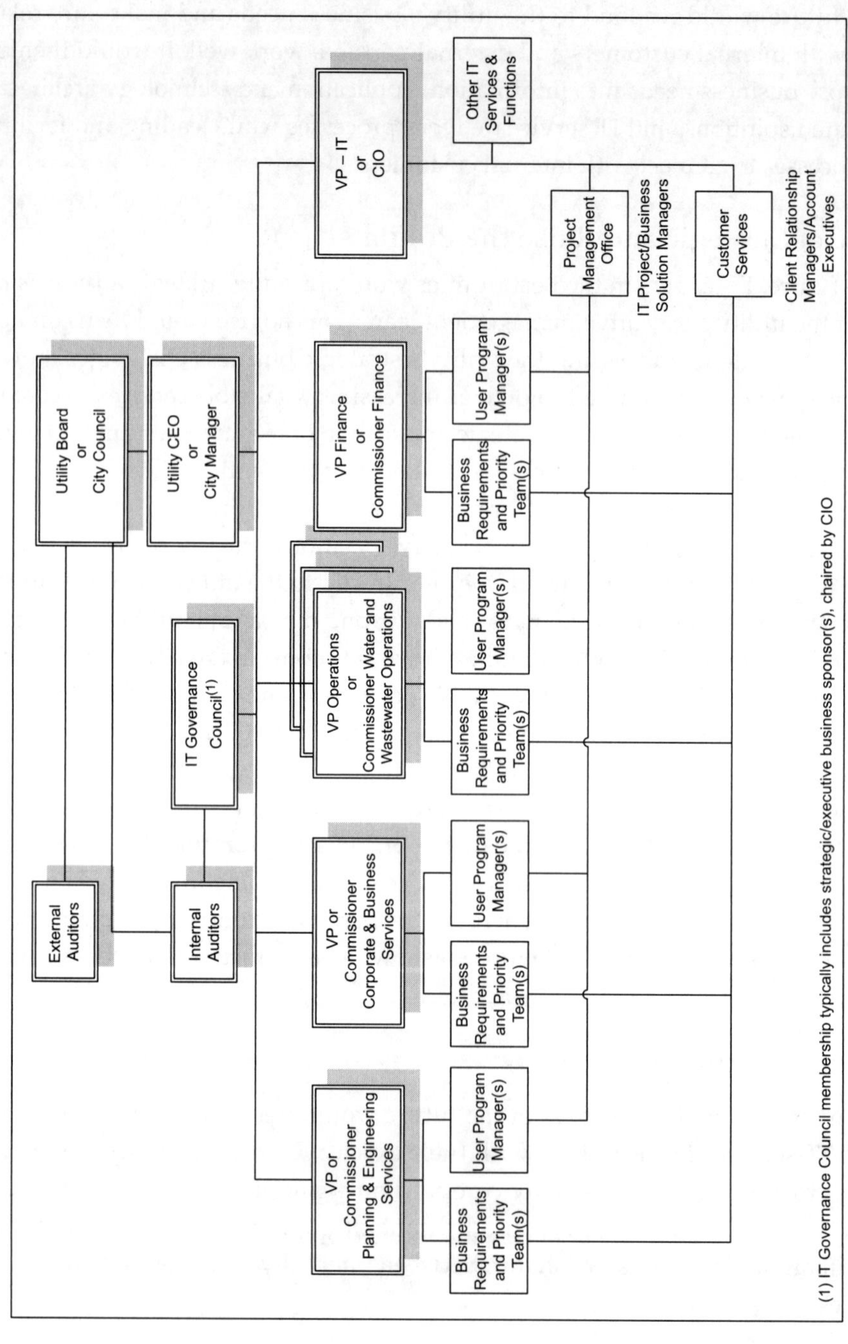

Figure 8.4 Best practice utility IT organization governance (Rau, 2004).

Requests for IT to provide leadership in management and integration of information systems solutions in response to business challenges and opportunities are being made with increasing frequency. This trend will continue in the next decade as technology is increasingly seen as a crucial tool in providing more value to customers and responding to an increasingly demanding regulatory environment.

In the future, some services might be provided by corporate IT; others by a specific department, nonutility front-line groups, or support-service groups; and yet others through external partnerships. These partnerships could include consultants, systems integrators, hardware maintenance providers, and SAAS providers. Some organizations, like the city of San Diego, have established separate corporations to deliver IT services to all departments, including the water utility.

Regardless of who is providing IT services, decision-making responsibility should be clearly reflected in the governance structure, the structure of the IT organization, and the overarching enterprise architecture. Weill and Ross (2005) presented an IT decision-making matrix that includes six types of governance archetypes, five IT decision-making domains, and the opportunity to assign decision-making or input rights to each domain per group or individual. This matrix can help utilities clearly assign responsibility for making key decisions relating to IT. The archetypes include

(1) *Business monarchy*, which is a centralized, business enterprise-centric model, where responsibility for IT decision making and resource allocation is assigned to senior business executive(s) or an IT governance council for one or more key IT decision-making domains.
(2) *Information technology monarchy*, which is a centralized, corporate IT-centric model, where responsibility for IT decision making and resource allocation is assigned to the CIO and IT leaders for one or more key IT decision making domains.
(3) *Federal*, which is a shared model, where responsibility for IT decision making and resource allocation is assigned to both front-line departments and the corporate IT department. Business-driven collaboration between the front-line departments and corporate IT drives the organization-wide optimization of IT and its role in the utility.
(4) *Information technology duopoly*, which is a shared model, where responsibility for IT decision making and resource allocation is assigned to department heads or commissioners and the CIO and IT leadership. Organization-wide optimization of IT is dependent on collaboration between the front-line departments and corporate IT.

(5) *Feudal*, which is a decentralized, business-unit-focused model, where responsibility for IT decision making and resource allocation is assigned to managers of business units or business processes.
(6) *Anarchy*, which is the most decentralized model, where responsibility for IT decision making and resource allocation is assigned to individuals or small groups.

The five decision-making domains are (1) IT principles and policies, which describe the high-level strategies that guide the way IT would provide most value to the utility; (2) IT architecture, which describes the overall framework for technologies, standards, and specifications that provides context for all IT systems; (3) IT infrastructure, which describes the specific hardware and communications infrastructure required to provide access and information sharing across the utility; (4) business application needs, which describe the business needs and related application software capabilities required to run the utility; and (5) IT investments, which describe how much money will be invested in which part of the service or organization. Each of these decision-making domains can be assigned to the enterprise, department, business unit, or group, and the individual levels.

Weill and Ross (2005) proposed that stakeholder groups or individuals be allocated either input or decision rights in each domain based on the aforementioned archetypes.

3.8 Manage the Change

Managing change well is the final step for utility IT organizations to undertake to ensure a successful transformation from the current as-is state to the target to-be maturity level, organizational structure, and governance framework. Although there are many ways to manage change, the process is often cut short for expediency reasons or lack of available change-management skills. A recent survey of the U.K. public sector highlights the shortage of change-management expertise as being the number one skills-related hurdle to improving municipal performance.

Successful changes are those that make certain all important change elements are cared for in a timely manner. A compelling vision would have been in place, a shared sense of urgency would have fueled the journey, visible leadership would have guided the process, a clear plan would have been followed, and the appropriate resources would have been applied along with incentives to deliver in a reasonable timeframe. The change equation in Figure 8.5 shows elements that govern a successful transformation. It is important to recognize that if any of the elements in the numerator in the equation is zero, there is no change possible.

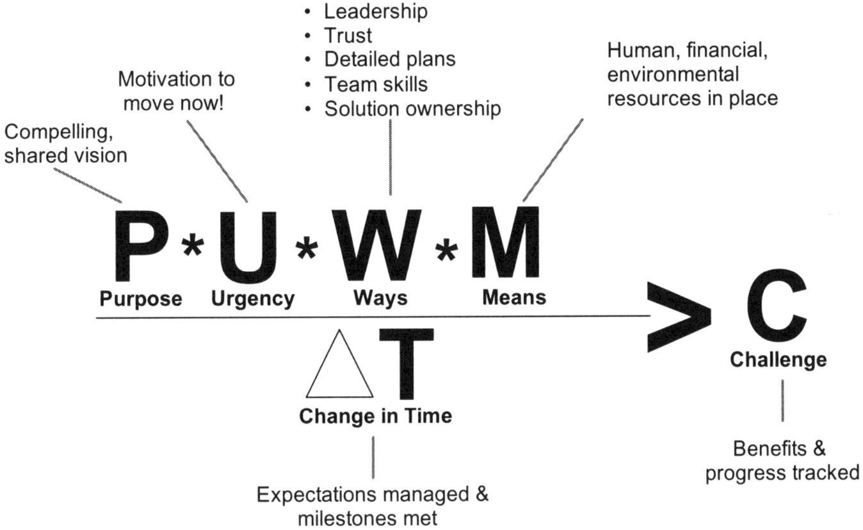

FIGURE 8.5 Change equation.

There are a number of examples of methodologies that can be used to guide the change process. Some find their roots at the personal-change level, like the best-selling *The 7 Habits of Highly Effective People* (Covey, 1989) and *ADKAR: A Model for Change in Business, Government and Our Community* (Hiatt, 2006). The former is based on the "maturity continuum" of personal and interpersonal effectiveness and on the development of habits created by combining knowledge, skills, and desire; the latter combines personal and professional awareness, desire, knowledge, ability, and reinforcement.

Other change models find their genesis in IT project methodologies such as Prince2, which uses the following 5-step process: (1) set a common purpose, (2) mobilize resources, (3) plan and design the solution, (4) implement the solutions, and (5) secure the result. Prince2 is a process-based approach for project management, providing an easily tailored and scalable project management methodology for the management of all types of projects. It does not spend much time on the underlying motivations of people, but is focused on delivering the objectives of the project. The method is the de-facto standard for project management in the United Kingdom and is practiced worldwide.

Yet other change models are aimed at transforming organizations based on vision, people, culture, and results. For example, *Leading Change* (Kotter, 1996) and the follow-up *The Heart of Change* (Kotter and Cohen, 2002) outline reasons why change processes fail. The books describe eight phases of change as follows: (1) establish a sense of urgency; (2) create a guiding coalition and guiding teams; (3) get the vision right; (4) communicate the vision for buy-in; (5) enable action and empower people to clear obstacles; (6) create short-term wins; (7) don't let up and keep moving; and (8) make it stick.

All change models can help and the best one should be selected based on the type of challenge the utility IT organization faces. Regardless of the chosen model, success requires a level of stakeholder engagement and communications commensurate with the nature of the challenge.

This engagement and communication should occur in a strategic manner, captured in a framework that poses the following questions:

- Which are the highest priority stakeholders and how much effect do they have on the success of the change?
- What are the needs and interests of the priority stakeholders related to this?
- What communication and engagement strategies should we follow for each stakeholder group?
- Want messages are appropriate for the group?
- What specific methods should we use to communicate?
- What specific events would be best and how frequently should they occur?
- Who should be responsible for communicating?

The final element to manage well in the change process is development of the culture. Often, there are opportunities for utility IT organizations to adjust their culture to better align with the culture of the overall utility organization. This can be done by basing behavior and decisions on a set of clear values and principles and becoming more customer-oriented and responsive. There are three powerful drivers that result in true culture change and transformation in utilities. These are

- The "burning platform" of fear, risk or disaster, such as in Milwaukee, Wisconsin, where a cryptosporidium outbreak in 1993 brought in the need for significantly improved quality monitoring and treatment management while putting the brakes on a trend to transform to "city-as-a-business" governance constructs that had been brought in by Mayor John Norquist in 1988;
- The "wolf-at-the-door" prospect of competition, such as the entry of private-sector water and wastewater operators into the North American marketplace in the late 1990s and early 2000s; and

- The "nirvana" of being the best, such as utilities in Colorado Springs, Colorado (e.g., Colorado Springs Utilities), and Edmonton, Alberta, Canada (e.g., EPCOR), which both quickly transformed from responding to competitive pressures to focusing on great results and having award-winning governance frameworks in place.

The first two drivers are more likely to provide the momentum for change as the lack of true carrots and sticks in the public sector make it difficult for leaders to create the necessary sense of urgency. The third driver is the aspirational motivator that delivers the most sustainable culture change, but is the most difficult to achieve.

The IT response to these cultural change drivers should be swift and accurate. If the utility's challenge is related to a number of environmental disasters caused by poor management of treatment systems and sampling information, IT should be there to help with technology solutions. This requires an IT culture that includes a mix of strategy, customer-service orientation, responsiveness, and utility business awareness. In addition, if the utility is under siege from a competitor, the IT organization needs to be sensitive to the effect of that situation on their customer staff. Chapter 9 provides more detail on the types of challenges that lie ahead.

The utility IT organization itself has a similar set of drivers. Off-shoring of services; availability of services through local consultants, contractors, and vendors; and the increase in the number of application services that can be provided outside the corporate firewalls all provide competitive challenges. These drivers can provide the impetus for IT to change its culture to be more flexible, business-oriented, and focused on the well-being of the utility. In addition, IT governance has an important role to play in setting the right philosophy and direction and managing integrated IT service delivery performance. The governance vision for leading utility IT organizations in the future should include recognition by industry associations.

4.0 REFERENCES

Covey, S. R. (1989) *The 7 Habits of Highly Effective People;* Simon & Schuster: New York.

Hiatt, J. M. (2006) *ADKAR: A Model for Change in Business, Government and Our Community;* Prosci Learning Center Publications: Loveland, Colorado.

Kotter, J. P. (1996) *Leading Change;* Harvard Business School Press: Watertown, Massachusetts.

Kotter, J. P.; Cohen, D. (2002) *The Heart of Change,* 1st ed.; Harvard Business School Press: Watertown, Massachusetts.

Rau, K. G. (2004) Effective Governance of IT: Design, Objectives, Roles and Relationships. *Inf. Syst. Manage.,* **21** (4), 35–42.

Salkowitz, R. (2008) *Generation Blend—Managing Across the Technology Age Gap;* Wiley & Sons: Hoboken, New Jersey.

Weill, P.; Ross, J. W. (2005) A Matrixed Approach to Designing IT Governance. *MIT Sloan Manage. Rev.,* **46** (2), 26–34.

5.0 SUGGESTED READINGS

Baldoni, J. (2003) *Great Communication Secrets of Great Leaders;* McGraw-Hill: New York.

Salkowitz, R. (2008) *Generation Blend—Managing Across the Technology Age Gap;* Wiley & Sons: Hoboken, New Jersey.

Shapiro, R. J. (2008) *Futurecast—How Superpowers, Populations, and Globalization Will Change the Way you Live and Work;* St. Martin's Press: New York.

Chapter 9

Critical Success Factors and Key Future Challenges for Information Technology in Water and Wastewater Utility Projects

1.0 INTRODUCTION

Information technology (IT) critical success factors are the most important elements of successfully defining, procuring, designing, and implementing IT projects. The success of IT projects is typically measured by the following basic project parameters:

- *Scope*—does the IT system meet the needs of the client?

- *Schedule*—is the system delivered on time?
- *Budget*—what will the IT system cost?
- *Staff*—what resources are available and should work on the project?
- *Metrics*—how will success be measured?

Success factors fall into the following categories: people, process, technology, and measurements. These factors are discussed in the following sections. Addressing critical success factors in a structured, organized, and consistent manner will help to achieve overall project success.

The final section of this chapter (Section 2.0, "Key Future Challenges") addresses the key future IT challenges facing water and wastewater utility operators today. As stated at the beginning of this manual, because IT is constantly evolving, the reader should keep in mind that the thoughts presented here are predictions.

1.1 People

Early involvement of end users of IT is essential as is the necessity to establish an effective method to manage their involvement. Therefore, project roles and responsibilities should be clearly defined. Understanding the ability of the organization to change and assimilate technology and having in place methods to obtain user acceptance are also important factors for implementing IT in water and wastewater utilities. Finally, a willingness to understand and accept change as scope evolves is also essential to IT in water and wastewater facilities.

1.2 Process

It is important to develop and implement a detailed, structured project management framework for the execution of IT projects. This framework includes the software development life cycle and a mature overall methodology for the delivery of systems.

Achieving the optimal balance of flexibility and structure in the methodology is also important. Early phases of IT projects (e.g., user requirements, technical requirements, and preliminary design) are extremely important because any errors made during these early stages are much more expensive to correct than the errors that happen later in the process (e.g., coding or implementation). In an ideal world, it would be best to make sure that all the preliminary tasks are fully completed so that the subsequent phases can build upon the previous work. Early methodologies emphasized this linear and unidirectional ("waterfall") approach, typically leading to disastrous results. Modern methodologies emphasize the iterative process, where each phase of the process is revisited and adjusted based on lessons learned.

It is important to understand the business reasons for implementation of IT systems, the business processes that are supported by IT, and the means to coordinate different groups in the utility to build and maintain IT systems so that the utility, as a whole, generates value. Utilities need a blueprint for IT, one that clearly shows how all the IT pieces fit together and the purposes they serve. These blueprints typically exist in a utility, but they may be loosely defined documents, driven by general policies, or they may exist only in the heads of experienced utility personnel. There are always opportunities for cost, performance, and quality enhancements using IT; however, they will only generate value when the utility has an architectural blueprint that binds business drivers to IT support systems. Therefore, IT project managers should consider development of an enterprise business model to bring rigor to IT implementation plans.

1.3 Technology

It is important to consider not only the needs of a specific project or application, but also the overall system integration to avoid "islands of information" or "information silos."

An IT project manager should be critical of the latest trends because sometimes they are simply older trends with a new label. Take advantage of technology that is specialized for utility users and the types of systems utilities manage. Do not expect every capability to come "out of the box," but do expect that customization services will be required. Understand the new software delivery models, such as software as a service and utility computing, as these have the potential for reducing IT personnel, licensing, and maintenance costs.

In addition, an IT project manager should not expect IT industry standards and open-source software to solve IT problems. Standards commoditize IT capabilities, a natural progression in technology development that levels the market playing field and puts the focus on generation of high-value applications, which is where good vendors (because they are profit-driven) place emphasis. As such, generating value to utility customers should be the focus of improvements in IT.

1.4 Measurements

A clearly defined target is easier to reach than an abstract foray in a general direction. In addition to critical success factors general to good project management, each project should specifically identify business process outcomes expected from the project. In too many cases, potential benefits can be missed unless the implementation team knows about the desired outcomes before implementation.

Identification of success metrics before implementation allows the utility to start measuring current state, allowing before-and-after comparisons. It is important to understand that an IT system that is no longer changing is no longer being used. Therefore, at what point will the original system implementation be "complete" and the system transition to "ongoing" enhancement mode? This is especially critical for IT projects. Reasonable requests for system enhancements are proof of end user engagement: Users can see how the new system can improve their business processes and suggest ways to continue to improve business processes. This highly desired state, however, is in direct opposition to the organization's need to "complete" the implementation project. Clearly defined targets will help keep the original implementation on track. A clearly defined "change request" process will capture legitimate enhancement requests and allow them to be prioritized and acted upon in a manner that best meets business objectives.

2.0 KEY FUTURE CHALLENGES

This section discusses the key future challenges for IT in water and wastewater utilities. Again, because IT is constantly changing, the reader should keep in mind that these challenges were predicted at the time the manual was published.

2.1 Electronically Stored Information

Utilities will need to respond to the new requirements related to legal discovery of electronically stored information (ESI). In December 2006, the Federal Rules of Civil Procedure were amended, thus increasing the pressure on private and public sector organizations to produce all relevant electronic data when requested during litigation. Some high-profile cases (with large penalties against companies that failed to provide ESI adequately) have also underscored the need for organizations to address these new and more stringent requirements. For example, the failure to properly collect and produce ESI during discovery directly led to a $28-million adverse jury verdict in the "simple" sexual harassment case of Zubulake vs UBS Warburg in 2003. In another case, Qualcomm Inc. vs Broadcom Corp. in 2008, a company was ordered to pay $8.5 million for "intentionally withholding tens of thousands of documents from its opponent in an effort to win this case." These cases underscore the need for water and wastewater utilities to enhance their ability to produce ESI in legal cases and in response to public information requests.

2.2 Capability Enhancements and Cost Reductions

Information technology changes are expected to benefit water and wastewater utilities through capability enhancements and cost reductions. Trends in utility computing (i.e., packaging of computing capability as a metered service), the Internet, "cloud computing" (i.e., using the Internet to distribute and scale software services), and software as a service will continue to drive down the cost of managing foundation IT elements and make it easier to obtain IT services. It may not be necessary to maintain a large staff for IT development and maintenance. Specialized IT systems will still have the highest cost; however, their method of delivery will change such that it will become unnecessary to purchase hardware and maintain software licenses.

2.3 New Applications for Information Technology

Information technology systems will continue to penetrate further into management areas of water and wastewater utilities. Technology applied to the management of engineered unit processes in water and wastewater utilities will find new application in managing the utility business. Business process automation will help to improve efficiency in utilities, including better systems for normal business functions, such as human resources and financials, and supervisory operations and integration across the utility enterprise so that, for example, real-time water demand and energy cost can immediately affect unit-process operating strategies and source water management or real-time water demand adjustment (actually affecting demand pattern) based on costs and other optimization criteria.

2.4 Increasing Cooperation across Departmental and Organizational Lines

As technology becomes more robust, the possibility of effectively "sharing" infrastructure increases. As technology becomes more complex, the desire to leverage scarce resources increases. As the cost of computing at a distance decreases, whether across town or across the world, options increase for collaboration. Utilities will be offered new opportunities and will need new criteria to evaluate risks and advantages. In some cases, such as human resources, outsourcing may provide much higher robustness than an individual utility could otherwise cost justify. Geographic information systems, based on shared maps and parcels, is already leading many utilities into previously unexplored areas of collaboration. Previous shared radio systems, which had insufficient capacity to meet all needs during a crisis event, forced utilities to develop separate infrastructure. As new infrastructures are put in place with adequate capacity and technology safeguards to support all users in crises situations, collaboration opportunities can be revisited.

3.0 REFERENCES

Qualcomm Inc. v. Broadcom Corp. Case No. 05cv1958 (BLM) 2008 U.S. Dist. LEXIS 911 (S.D. Cal. Jan 7, 2008) (*dismissed in part and remanded on different grounds).*

Zubulake v. UBS Warburg LLC, 217 F.R.D. 309, 312 (S.D.N.Y. 2003).

Chapter 10

Case Studies

(*continued*)

1.0 CASE STUDY 1: A SOUTHEASTERN U.S. AUTHORITY

1.1 Introduction

A medium-size utility (hereinafter referred to as the *authority*) in the southeastern region of the United States conducted an information technology (IT) strategic plan subsequent to their establishment as an entity separate from their (previous) parent municipality. The authority serves 73,000 water, 71,000 wastewater, and 10,000 reclaimed water customers, and owns and operates 20 water plants and 10 wastewater plants, maintaining 1890 km (1174 miles) of water mains, 1575 km (980 miles) of wastewater mains, 390 km (242 miles) of reclaimed water mains, and 357 wastewater pumping stations. With a 165-person workforce, the authority treats and distributes approximately 133 ML/d (35 mgd) of potable water and reclaims 80 ML/d (21 mgd).

At the outset of the planning process, the authority was acquiring IT services from the IT department of the municipality of which they had been a part under a service-level agreement based primarily on the previous interdepartmental method of distributing costs within the municipality. Whereas many utilities consider the possibility of outsourcing some or all of their IT support, incorporation of the authority put the utility in the opposite position where they suddenly needed to determine which, if any, IT services should be brought in-house. The purpose of the plan was as follows:

- Develop a detailed understanding of the services provided by the municipal IT department (hereinafter referred to as the *IT group*), including understanding what the IT group did well; which authority IT needs, if any, were outside of the core focus of the IT group; and whether and/or how the IT group support varied by physical location, area of service, and service volume level.

- Ensure the authority's significant financial investments in IT are being leveraged appropriately.
- Identify IT risks and opportunities and ways to address these issues.

1.2 Vision for Information Technology

The first step in the plan was to document current applications in use at the authority, most of which would require version upgrades within the next 5 years. From the beginning of the project, both the authority and the IT Group believed additional resources were required to support the Authority based on their rapid expansion and growth.

The vision statement for the authority's IT evolved through a series of workshops and detailed interviews with personnel to read as follows: The authority will provide its users and customers reliable access to the information needed for excellent service, tracking assets and performance, and promoting informed decision making by using the most effective internal and external technology support alternatives available.

1.3 Information Technology Business Practice Assessment

The purpose of this tool is to focus on how IT is selected, implemented, used, and managed within the authority. This tool allows utilities to compare their internal practices to other utilities within the industry who participated in the American Water Works Association Research Foundation (Denver, Colorado) "Creating Effective Information Technology Solutions" project. For the authority, it also illustrated the differences in perspective and practice between IT services and practices provided by the IT group, in general, and to the authority, in particular. To achieve this perspective, internal authority staff completed the assessment separately from the IT group staff. The two groups then came together and reached a consensus on the level appropriate to the authority.

Survey outcomes included the following:

- An awareness by the survey team of what other water utilities are doing within IT;
- Increased awareness by the authority of the best practices currently being performed by the IT group;

- Highlighted authority and IT group strengths compared to other water utilities; and
- Identification of areas for improvement within the authority compared to other water utilities.

The reader should note that IT group security practices were at industry best-practice levels. As a result, few specific recommendations related to security were required in the plan.

1.4 Data Quality Assessment

The authority's application status is unique in that one of the major applications, the customer information system, is owned and operated by others. A work management system is in early implementation stages and the financial information system is about to be replaced. Although data have existed in more than one application for some time, in the past this was primarily out of the authority's control. Moving forward, however, this situation is changing in the authority's favor. For now, automated data transfers, typically referred to as *interfaces*, are a future goal. Current opportunities are related to "practices" and how the authority can leverage existing resources to improve information quality in the various applications now available to them.

The authority's data quality assessment provided a structured method of identifying business processes producing poor quality data. Assessment recommendations included the following:

- Additional IT project management to ensure that IT applications are implemented in a way that leverages the investment to benefit the authority without imposing unnecessary requirements;
- A comprehensive review of document management processes;
- Researching an automated, secured, and Web-enabled permit tracking system; and
- Establishment of a data stewardship program focused on groups of data across all appropriate applications to develop trackable procedures to improve data quality. Areas needing immediate attention include meters, field assets, plant assets, and personnel.

1.5 Current State of Technology

A detailed IT user survey was conducted to obtain information on the current state of IT at the authority including the level of employees' technological skills and their varying technology needs. The survey identified 17 key applications and a general desire by most users to learn more about these applications. Overall results, outlined as follows, were mixed:

- Simpler applications fared better;
- More sophisticated applications did not score as well;
- New applications scored surprisingly well; and
- A "tendency" to better support building with IT, although the best support actually went to the two groups with the most proactive users and the worst support went to groups in which the users were hardest to contact.

The plan recommended that the authority prioritize training where the desire to learn is high, where there are few users able to assist others, and where there is a high level of user dissatisfaction with available support. It is important to note that users identified "excellent" but not "adequate" resources available.

1.6 Organization and Governance Assessment

This task reviewed the current organizational structure and proposed various IT organization alternatives, both in structure and timing. Discussion included staffing levels and reporting structures and policies and service-level agreements (SLAs).

1.7 Service-Level Agreement and Service Catalog

A draft SLA formalized the arrangement between the authority and the IT group (the service provider) to deliver identified support services at prescribed levels of support and at an agreed-upon cost. Subsequent to completion of the plan, this document was reviewed by the legal department before being finalized by the parties. The SLA will evolve over time with additional knowledge of authority requirements and the introduction of new applications and services into the support portfolio provided to the authority. It was intended to complement current procedures, not override them. In addition, it was designed to encourage a joint, open partnership approach with regular and free exchange of information between both parties.

The service catalog was developed identifying services in use by the authority and detailing how these services would be provided. The authority obtains IT services through a variety of providers, including employees, vendors, software as a service (SAAS), and outsourced business services based on applications owned and maintained by business service providers. In most cases, the IT group acts as a single point of contact for IT-related services. In limited situations, however, it is more reasonable for authority employees to go directly to another point of contact. The service catalog was intended to document which services are and are not supported by the IT group. The service catalog captured the following details for IT services used by the authority:

- Service name,
- Tracking system,
- First-level responder,
- Second-level responder,
- Escalation contact,
- External support agreements,
- Power users,
- Business effect of loss of service (for key services only),
- Supported life cycle (for key hardware only), and
- Service availability target (for key services only).

1.8 Policies and Procedures

With separation from the municipality came a need for the authority to develop policies to replace those currently in place at a municipal level. While acquiring services from the IT group, it was appropriate to follow the policies in force for other uses of IT group services as closely as possible. Where the policies did not apply, new policies had not yet been developed. The plan proposed a general approach to policy development as follows: Review existing municipal policies and confirm or modify as appropriate for the authority and develop policies specific to the authority where required.

Policies and procedures help employees perform nonroutine tasks in a successful manner. They can provide extra benefit to an organization, such as the authority, during a time of transition. Policies and procedures should be short and contain as little boilerplate as possible. Suggested initial policies and procedures include the following:

- Who to notify and how such notification should be provided for scheduled service interruptions. This policy would need to be updated each time a group of employees was moved to a new location (a frequent occurrence during transition periods).

- Who to notify and how such notification should be provided if a specific service was causing a network problem. For example, if a printer was causing excessive network traffic, the supervisor of that area should be contacted before disconnecting the printer so that alternate printing mechanisms could be put in place.
- Who to notify and how such notification should be provided when a broadly used system resource becomes unavailable. For example, an e-mail is sent out when a specific application went down, but this e-mail could not be received until the application itself was back online. Text messaging the cell phones of users who are dependent on that application could minimize their time spent trying to regain system access.

1.9 Information Technology Roadmap and Final Information Technology Strategic Plan

Based on findings in the previous tasks, an IT roadmap was created highlighting project tasks and presenting a preliminary recommended path forward. In summary, the IT strategic plan outlined the systems, applications, practice improvements, organizational support, recommended sequence, and resulting costs. It also included creation of a service catalog and subsequent SLA to formalize arrangements between the authority and the IT group.

2.0 CASE STUDY 2: METROPOLITAN WATER DISTRICT OF SOUTHERN CALIFORNIA

Metropolitan Water District of Southern California (MWD) (Los Angeles, California) is a consortium of 26 cities and water districts that supplies water to nearly 18 million people. The MWD currently delivers water at an average of 75 m^3/s (1.7 bil. gal/d) to a 13 470-km^2 (5200-sq mi) service area. The MWD has two main water resources, the State Water Project and Colorado River Aqueduct, which are capable of providing 83 and 26 m^3/s (1.9 and 0.6 bil. gal/d), respectively.

In 2005, MWD embarked on a project to develop a mathematical model-based decision system to support its planning and operations. The system included a number of stakeholders from different parts of the organization, and each group (e.g., operations, engineering, and planning) had different focuses and objectives.

Gathering information on user requirements started by defining the use cases. This process included the following steps:

- Interviews with stakeholders.
- Defining use cases
  a) Entering data about use cases onto a Web site where use cases were hierarchically organized,
  b) Based on use cases, the team identified a list of specific functions that the system would need to perform, and
  c) The team developed a structured acceptance testing procedure based on use cases and the list of functions.

2.1 Initial Interviews with Stakeholders

Interviews were conducted using prepared questions that would solicit information that would be required for defining use cases. In the beginning, a general template for interviews was developed, but it was adjusted somewhat after each interview to take advantage of lessons learned in earlier interviews. Initial interviews were conducted separately with different stakeholders.

2.2 Defining Use Cases

Interview notes were used to prepare use cases by filling in a template. An example of a use case template filled in for a specific use case is provided in Table 10.1.

Draft use cases were reviewed for accuracy by stakeholders.

2.3 Entering Data onto the Web Site

Data from the forms was entered onto a Web site that was placed within the MWD Intranet (inside the firewall) and made accessible to all team members. Figure 10.1 shows a partial list of the use cases, organized hierarchically in parent-child structure.

The lower part of this Web site was allocated for defining the data used by each use case; however, because of budgetary constraints, user requirements did not go to this level of detail.

2.4 Creating a List of Specific Functions

Based on the use cases, a list of specific functions was defined. An excerpt from this list is shown in Figure 10.2.

TABLE 10.1 Example of a completed template for a use case.

Business process/ decision elements	Explanation	Entry
Business process_ name	The name of the business process or software function	Add_new_service_connection
Business process_ objective	The objective of this business process/ software function	To provide a new connection to a member agency
Business process_ description	A narrative description of this business process/software function; list the steps that are included in this business process and also list actions that are taken	Steps include determine the best location for the connection; examine what the pipe can deliver (capacity) at the connection point; examine if the member agency's system can take the flows; and if there is a match, design the new connection
Trigger	What initiates this business process and/ or demands this functionality?	Member agency request
Timing	How often does this business process occur?	Occasionally, according to trigger
Decision_level	What type of a decision/business process is this?	Tactical
Actor	Who is the primary person responsible for this business process/decision?	Operations
Business process _measures	What is the measurement used to evaluate this business process?	Design a turnout that will deliver requested flow and requested pressure range; maintain pressure in the main line
Business process _target	The target performance measurement of this business process.	Meet required flow volume at requested location
Business process _parent_names	If this decision/business process is part (or one "path") of other (higher-level) decisions, put the names of those higher decisions here.	Design new facilities
Business process _category	What is the category of this business process?	Operations and maintenance
Support system	Is this business process/decision currently supported by another specific software package?	Currently using steady-state model for capacity analysis and surge analysis model

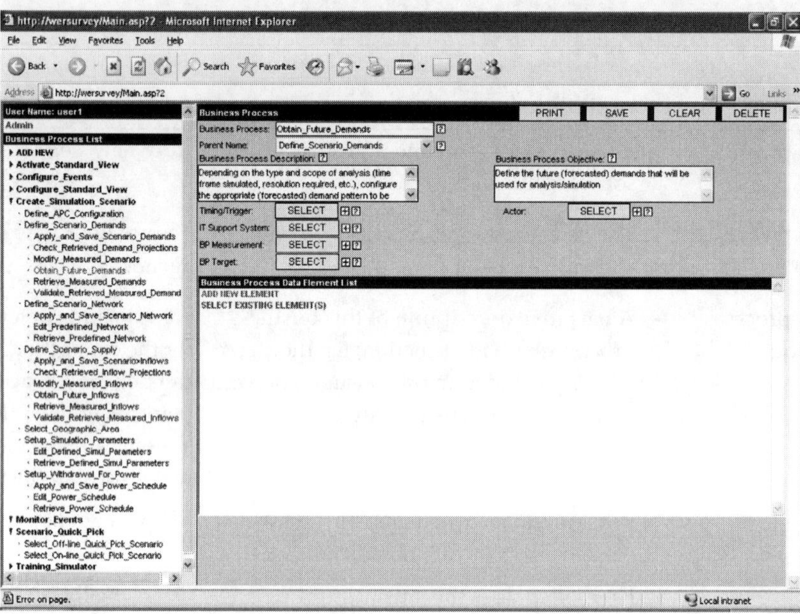

FIGURE 10.1 Web site screen shot showing some use cases.

Description of main functionality items	More detailed functionality	Critical?	Priority
Compute pressures and flows throughout the network using dynamic simulation		Yes	1
	Obtain steady state by executing the dynamic model until it converges to results for steady-state conditions	No	2
	Simulation of flow-through slide gates including the possibility to simulate the effect of throttling a gate	Yes	3

FIGURE 10.2 Snapshot from the requirements list.

2.5 Developing Structured Acceptance Testing Procedures

It is not practical to test each of the numerous functionality items by themselves; therefore, acceptance testing procedures included sets of different functions. An example of one acceptance testing procedure is shown in Figure 10.3.

2.6 Summary

The purpose of user requirements is to

- Provide an efficient mechanism for communication between users and technologists and
- To provide a mechanism for users to verify that the system is indeed performing as they specified.

Narrative descriptions were not deemed optimal for achieving the main goals of the project. The project emphasized diagrams and structured lists and then linked the description of the desired functionality to a procedure that could be used for user acceptance. Therefore, the process provided a continuous link between different phases of user requirements development, from stakeholder interviews to acceptance testing.

7	Demonstrate the ability of users to define scheduled scenario simulations using a simplified interface. This test shall validate item 55.	Full Pilot Model	0.5 hours			
				1	Open the custom application (RTOS Scenario Manager) for editing scenarios.	
				2	Select one of the existing scenarios in the tree view on the left.	
				3	Select the "Schedule" tab.	
				4	Define a scheduled time for the scenario simulation to execute.	
				5	Verify on the simulation server that the scenario actually executes at the specified sheduled by checking the Windows event log.	55

FIGURE 10.3 Example of acceptance testing procedure that includes several functions.

3.0 CASE STUDY 3: HARDWARE AND SOFTWARE ARCHITECTURE AT JEA, JACKSONVILLE, FLORIDA

JEA is the water, wastewater, and electric utility in Jacksonville, Florida, that provides water service to more than 250,000 customers, with 100% of the supply coming from groundwater sources. Water supply comes from 32 well fields and associated treatment facilities. The distribution network is divided into two major grids encompassing a four-county service area with 2,800 miles of water lines. JEA has been challenged to reduce withdrawal of water from wells to meet tighter consumptive-use permit limits while raising the water quality and lowering total operating costs.

JEA has implemented an automated software system for water supply and distribution that significantly improves water operations. The system implements a rules-driven optimization that both proactively plans for, and reactively responds to, dynamically changing consumption and other water system changes, including daily changes in energy pricing. This system, called *water operations optimization* (WOO), uses a supervisory control and data acquisition (SCADA) system and data collected from a number of other sources and minimizes cost while improving the operating performance of the entire water system. The system manages JEA's consumptive use of local water resources, controls and monitors water quality in real-time, maximizes the value of energy for JEA's electric utility, and maximizes the existing capacity of water system assets to defer or eliminate the capital cost for new infrastructure.

The JEA adopted an architecture with key features recommended by the ARC Advisory Group, as shown in Figure 10.4. Field process control devices interface with a SCADA system that manages essential controls. The WOO application provides supervisory control. Water operations optimization determines well-pump and high-service-pump schedules, manages water quality, enforces work processes, generates alerts for crucial conditions, and trains operations personnel. The design includes "fail-over" and fail-safe mechanisms for fault tolerance, interfaces to external systems such as Web applications and mobile devices, and integration with business systems for exchange of management metrics used to continually assess the performance.

JEA uses a real-time modeling and reasoning technology platform to launch this application. The general architecture for model-based decision support is shown in Figure 10.5. Techniques applied include rule-based inference, neural networks, nonlinear constrained optimization, and mechanistic hydraulic and mass-balance water supply and distribution models. The real-time rules-driven platform integrates and coordinates these components in a single environment. Water operations

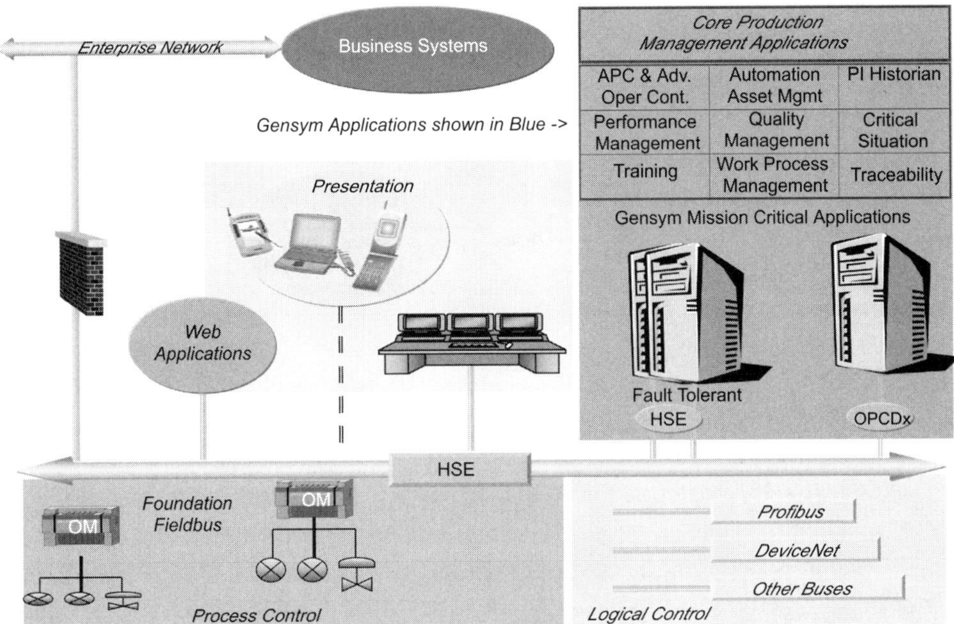

FIGURE 10.4 Architectural components in the JEA application (adapted from ARC Advisory Group).

optimization provides event detection and condition diagnosis capabilities that significantly improve the stability and reliability of WOO.

The architecture depicted in Figure 10.5 includes models for both water consumption (i.e., demand) and supply of water (i.e., well-field supply and distribution network). Ensemble neural network technology is applied to develop a nonlinear, adaptive consumption forecaster that is capable of retraining when real process conditions change. A hydraulic model of the distribution system was developed and then reduced in complexity to enable application in real-time automation. A mass-balance model was developed for use in open-loop decision support as a means of validating the plans and schedules determined automatically by the optimizer. Constrained, linear optimization is used to develop pump on/off schedules. Finally, event detection and condition diagnosis techniques were applied to ensure data quality and to proactively alert operations when important states or conditions occurred that may require operator attention.

The detailed functional architecture for WOO is shown in Figure 10.6 (Barnett et al., 2004; Jentgen et al., 2005). The managed physical assets include the well pumps in each well field, water treatment plant processes, reservoirs, and high-pressure

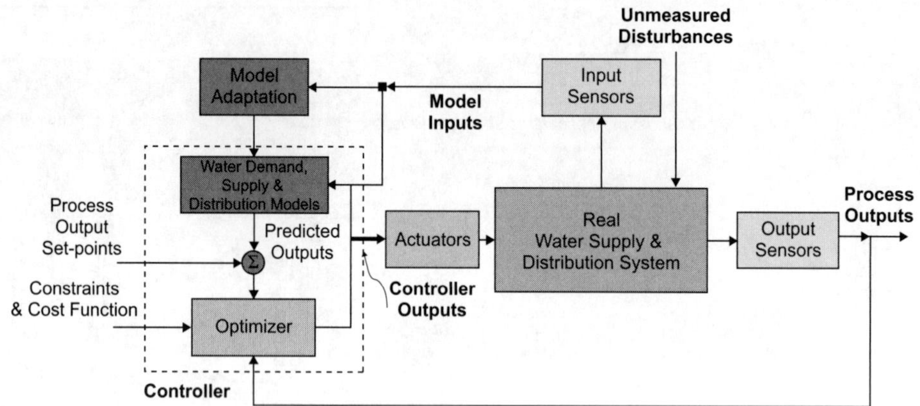

FIGURE 10.5 General architecture for model-based decision support.

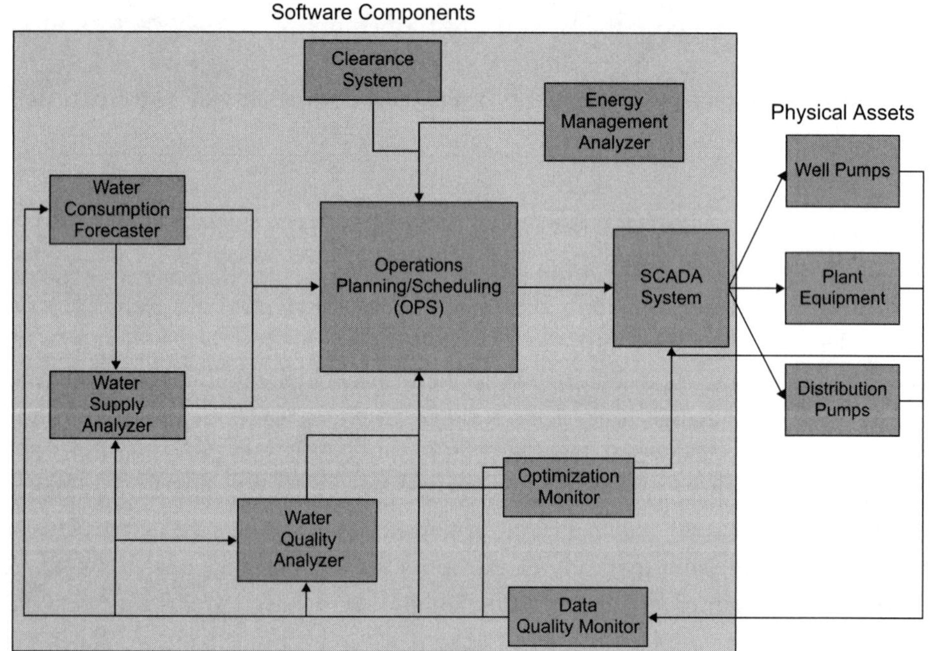

FIGURE 10.6 Water operations optimization functional modules (Barnett et al., 2004).

distribution pumps. With WOO, these assets are optimally controlled through the existing SCADA system. Software components include the following:

- *Supervisory control and data acquisition system*—direct interface to sensors, programmable logic controllers (PLCs), and human-machine interface.
- *Operations planner and scheduler*—applies hydraulic models and constrained optimization to allocate demand amongst multiple plants supplying the grid and then develops schedules for high-service pump operations from each plant.
- *Water consumption forecaster*—develops water demand profiles for virtual consumption points in the distribution network applying ensemble neural network technology. The water consumption Forecaster includes an adaptive feature that enables it to automatically retrain when conditions change in the water distribution system. The water consumption forecaster achieves greater than 90% daily accuracy over extended periods since it was implemented in 2003.
- *Water supply analyzer*—for each water plant (source of supply), the water supply analyzer applies models and constrained optimization to develop well-pump schedules for the groundwater sources.
- *Clearance system*—implements the approved process for identifying when equipment (e.g., pumps, valves, and chemical feed equipment) is removed from operation.
- *Water quality analyzer*—accepts real-time input from a number of sources, including field operators, SCADA, and laboratory information management systems, and uses this input to develop water quality operating parameters. The water quality analyzer proactively alerts system operators if there is an anticipated or actual water quality excursion in the system.
- *Energy management analyzer*—provides input to the operations planner and scheduler, including constraints such as a daily energy cost profile to enable scheduling of energy consumption that minimizes cost and maximizes the value of JEA generation during on-peak periods. This analyzer is the interface to JEA's electric utility and, in the future, can provide data from water facilities for distribution-system fault analysis and load studies.
- *Data quality monitor*—ensures the integrity of data needed by the WOO to evaluate plans, schedules, and controls. Includes smoothing, projection, and filtering combined with data substitution in cases where data points are missing or judged to be incorrect.

- *Optimization monitor*—compares actual system conditions to the forecast. When actual consumption varies significantly from forecasted consumption or key equipment failures occurring during the day, Optimization Monitor will alert operations and can force both a re-forecast of consumption and a re-plan of well-pump and high-service pump schedules.

Mapping of software applications to machines in the JEA network is shown in the architecture of Figure 10.7. This figure is used along with a "preflight" checklist to

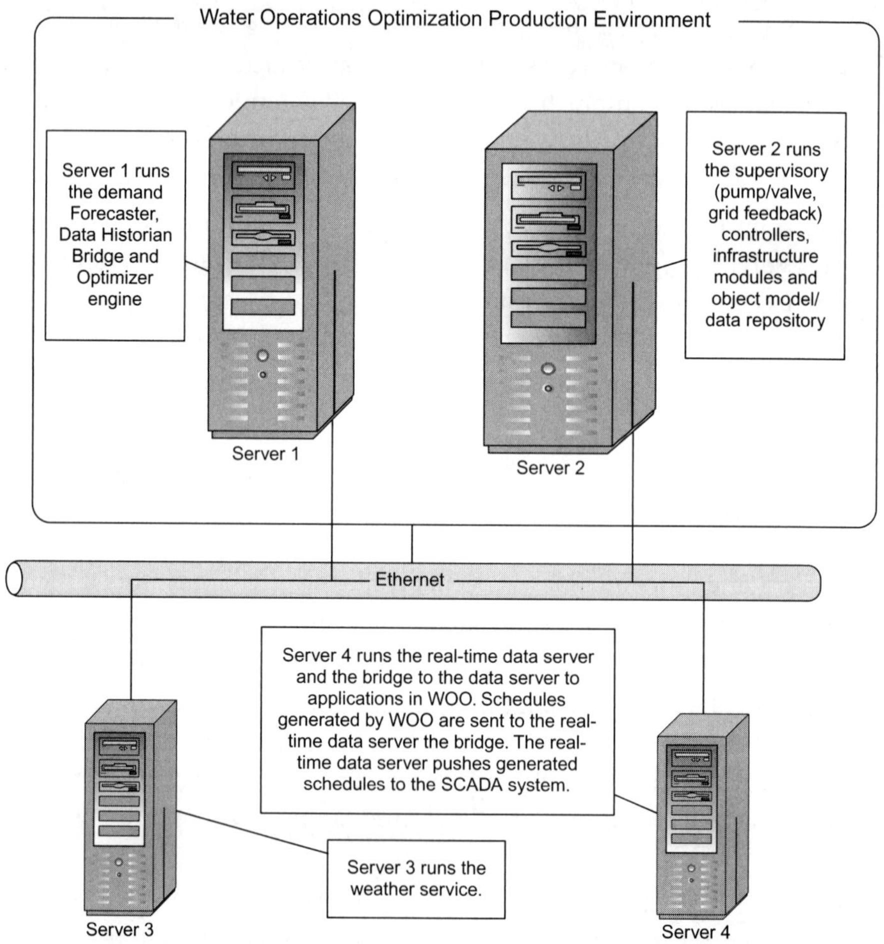

FIGURE 10.7 Water operations optimization hardware and software architecture.

identify each application, its executing condition, and any notes regarding the application (e.g., machine cycles used; random access memory, or RAM, used; planned maintenance; etc.) that may affect the overall up-time of the WOO system.

4.0 CASE STUDY 4: ENTERPRISE RESOURCE PLANNING SYSTEM IMPLEMENTATION AT ELSINORE VALLEY MUNICIPAL WATER DISTRICT, LAKE ELSINORE, CALIFORNIA

Elsinore Valley Municipal Water District (hereinafter referred to as the *district*) is a mid-size water utility that provides service to over 133,000 water, wastewater, and agricultural customers in a 250-km^2 (96-sq mi) service area in Western Riverside County, California. Beginning in 2005, the district's IT director realized that a large problem was looming on the horizon. The district's financial systems, including all major accounting, budgeting, human resources, payroll, customer service, and utility billing functions, were nearing the end of their useful life. The existing system was built as an entirely custom program, running on an outdated, 20-year-old software platform, and the vendor that supported it consisted of one person who was nearing retirement age. To address these issues and to gain other benefits from improved information systems, the district spent the next 4 years in a systematic process of developing an IT master plan, conducting an enterprise resource planning (ERP) requirements definition and system selection project, and implementing new financial information systems (FIS), human resources, and customer information systems (CIS).

4.1 Information Technology Master Plan

During a 12-month period, the district conducted an assessment of its current and anticipated data needs, identified opportunities to improve work processes, and planned improvements in the use of technology for the district. The results of the project were documented in an IT master plan that was presented to the board of directors in 2006. The plan outlined a series of projects over 5 years to significantly upgrade existing information systems to provide substantial, lasting benefits to the district.

The significant findings of the IT plan included the organizational need to access vital financial, billing, customer service, and water production data. The data were difficult or time-consuming to access, which inhibited management-level analysis and decision making. Business processes that were identified to have the highest potential for improvement included materials and inventory control, purchasing and payables, and capital improvement project (CIP) procurement and administration. A functional and technical fitness assessment of the district's existing information systems was also conducted, and the systems with the lowest fitness included financial, human resources, customer information and billing, and CIP management systems. An example of the resulting fitness assessment for the financial system is presented in Figure 10.8.

In alignment with these findings, the most significant recommended information system improvement was identified as a replacement of the district's financial, human resources, and customer service systems. These essential, core systems were based on custom software that had served the district well in the past. However, because of limited support and outdated technology, these systems inhibited the district from moving forward with greater efficiency and more effective business operations. The IT master plan recommended a series of projects for system replacement, under an overall program of "Financial, Customer Service, Human Resources Systems," that included the following:

- Financial information systems/CIS/human resources replacement program management,
- Integrated FIS/CIS procurement,

	Criteria Component	Assessment	Overall
Functional	Business Objectives & Vision	◑	◕
	Work Process	◕	
	Adaptable to Process Change	●	
	Features	◑	
	Productive to Use	◕	
Technical	Common Operating Environment	●	◕
	Information Systems Architecture	◕	
	Vendor Viability	●	
	Vendor Support	◕	

Low ← → High
● ◕ ◑ ◔ ○

FIGURE 10.8 Financial system fitness assessment summary chart (courtesy of Elsinore Valley Municipal Water District).

- Financial information systems implementation,
- Human resources procurement,
- Human resources implementation,
- Water production data warehouse, and
- Customer information systems/billing implementation.

The total budget allocated for these projects was $1.73 million over a 30-month timeframe that included requirements definition, software selection, implementation, and post-implementation support. The district proceeded with the first two projects, "FIS/CIS/HR Replacement Program Management" and "Integrated FIS/CIS Procurement," issuing a request for proposals (RFP) for consulting assistance to oversee and manage all aspects of replacement of the systems. The project also included development of user and technical requirements for FIS, CIS, and human resources systems, and assistance with selecting the appropriate vendors and system integrators. To reflect the encompassing role of the multiple systems involved, the project was renamed the "Enterprise Resource Planning Systems Replacement Project."

4.2 Requirements Definition and Systems Selection

The overall process used for ERP requirements definition and systems selection is shown in Figure 10.9. The first step in the process involved the review of business processes and defining the detailed requirements for the ERP systems. The district's consultant who was hired to do this work coordinated multiple rounds of interviews with functional groups of staff to develop detailed lists of requirements for the various functional modules of the ERP and the technical requirements. Template requirements were provided by the consultant, which were then reviewed and edited by district staff to develop a complete customized list of desired functionality. Ten functional areas were defined for the financial system, including general ledger, budgeting and planning, accounts payable, and accounts receivable; eight functional areas were defined for human resources, including applicant racking, benefits administration, payroll, and time and attendance; and 11 functional areas were defined for the CIS, including utility billing, payments processing, rates and fees, and meter reading.

On a simultaneous track to that of the requirements definition, the consultant conducted workshops to define "as-is" and "to-be" business process maps. A total of 20 business process maps were defined for the business processes identified by the IT master plan that had the greatest need for improvement. The process maps were

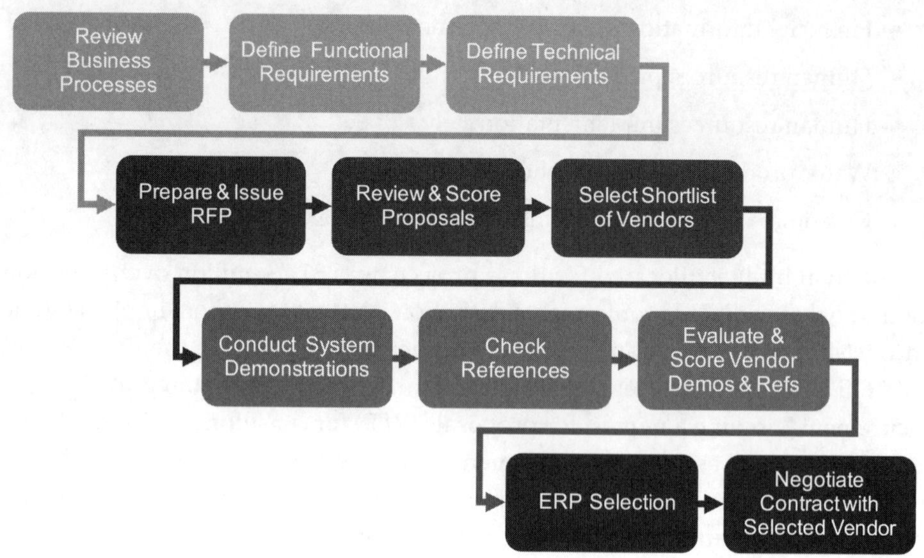

Figure 10.9 Enterprise resource planning requirements definitions and systems selection process.

developed during the workshops using a basic template in Microsoft Visio (Microsoft Corporation, Redmond, Washington) that allowed district staff to collaborate with the consultant in their definition. An example of one of the process maps is shown in Figure 10.10.

In the next step, an RFP was developed that contained all of the system requirements and business process maps that had been gathered by the consultant and the district for the ERP replacement project. The RFP was sent to 25 ERP vendors, some with overall integrated financial, human resources, and CIS, and others with one or more of the major system components. Six proposals were received in response including AMX International (Rexberg, Idaho) with Oracle JD Edwards EnterpriseOne; Cogsdale Corporation (Charlottetown, PE, Canada) with Microsoft Dynamics GP; Harris Computer Systems (Ottawa, Ontario, Canada) with GEMS NorthStar; Lawson Software (St. Paul, Minnesota) and Advanced Utility Systems (Toronto, Ontario, Canada); Tyler Technologies (Dallas, Texas) with MUNIS; and Wipro Technologies (Bangalore, India)with SAP. Total cost estimates for these proposals ranged from a low of $600K for Cogsdale and Microsoft Dynamics GP, to a high of $3.7 million for WiPro and SAP.

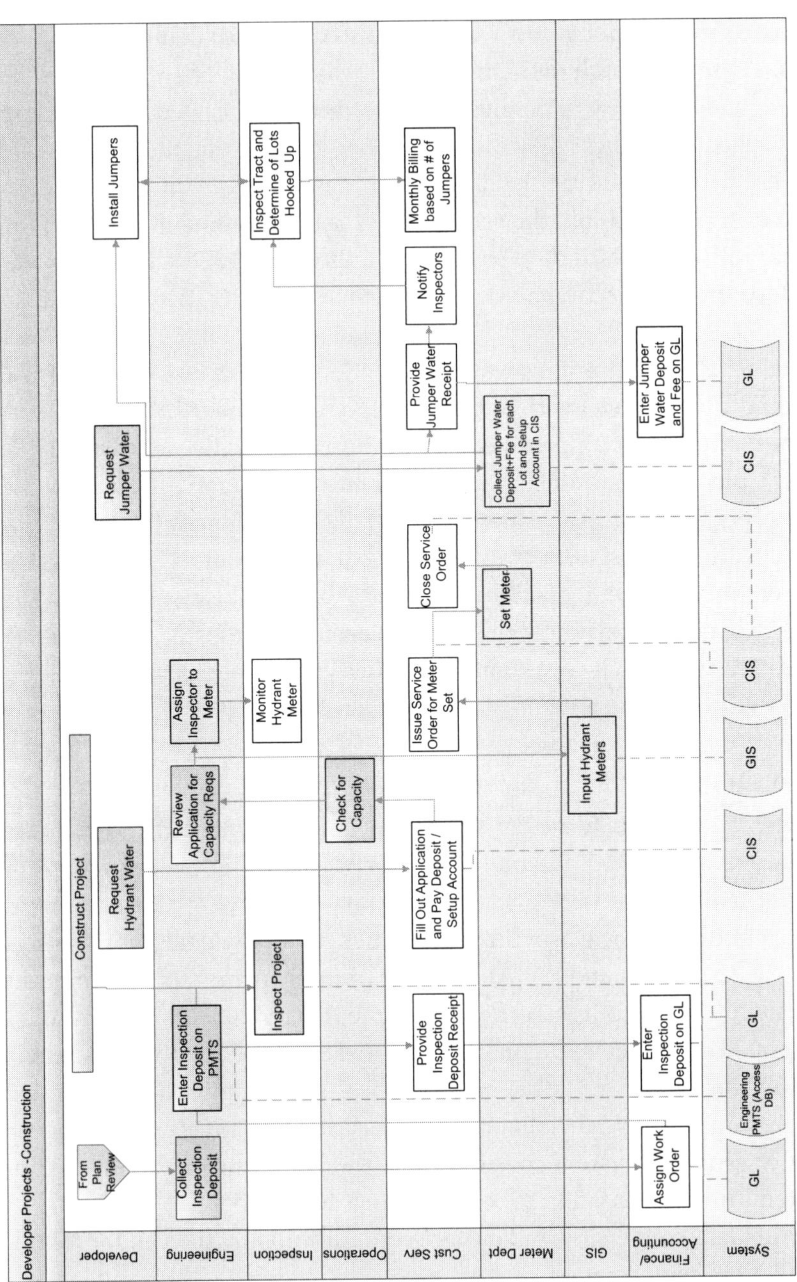

FIGURE 10.10 Example of "as-is" business process map (courtesy of Elsinore Valley Municipal Water District).

After review of the proposals from a qualifications standpoint and then with respect to cost, three vendor teams were shortlisted to conduct interviews and demonstrations. These firms included Lawson and Advanced Utility Systems, Cogsdale Corporation, and Tyler Technologies. The interviews consisted of 3 days of demonstrations based on a script that was developed by the consultant and sent to the vendors in advance. In general, the demonstrations were scheduled for the first day of a system overview and FIS, the second day for the remainder of FIS and human resources payroll, and a third day for CIS and utility billing. District staff from many different departments participated in all of the demonstrations and the scoring process.

The final result of the process was the selection of the team of Lawson Software and Advanced Utility Systems to implement the replacement ERP systems. After a lengthy contract negotiation process with both firms, the district's board was presented with a plan to replace the financial and customer service systems at a cost of approximately $3.6 million over the next 18 months. Although the board was surprised at a cost that was significantly higher than what had been estimated during the IT master plan, and nearly 50% higher than what had been included in the original vendor proposals, they knew that the replacement was needed and believed in the benefits of a system that would make the district more efficient and effective and enhance service to their customers.

4.3 Implementation

The implementation phase turned out to be much more difficult than what was originally estimated or expected. Despite some minor glitches in the contract terms and support agreements, both the financial system and customer service system vendors were able to get things started on time. The Lawson implementation was scheduled to "go live" at 12 months and the Advanced Utility Systems started 3 months later and was scheduled for go live at the same time in July 2009, the beginning of the new fiscal year. The implementation process for each of the vendors was remarkably different. Lawson's standard implementation was more traditional in that it started with training and software installation, followed by a discovery and design phase, and finally ended with system configuration and data conversion. The Advanced Utility Systems' implementation process was nearly the opposite, beginning instead with an offsite data-conversion and system configuration by the vendor followed by onsite installation, configuration, and training.

Even as months passed through the requirements definition, discovery, and design phases, district staff, who were involved throughout the process, were pleased with the progress being made. However, almost 2 months before going live, the project team realized how much work was left to be done and how many tough decisions needed to made. Unfortunately, many of the items discovered in the final months required multiple cycles of rework, more data conversion, and more difficult decisions to resolve them. Although some of the issues were actual limitations of the software, many had to do with the limited remaining time and overwhelming amount of new capabilities or the complexities of the existing custom-built system and difficulties in translating the hard-coded business rules. The result was that instead of steadily reducing the number of problems as go-live approached, the action-item list grew longer, and the effort required to surmount them stretched the implementation project team to their capacity.

After the district realized they would not be able to meet the original schedule of either vendor, the financial system go-live was delayed 2 months as the action items were reprioritized and concentrated efforts were applied to the final steps of implementation. A similar situation in the customer service/utility billing software implementation resulted in a 6-month delay for go-live. Together, the implementation delays caused an increase in project costs of approximately $600,000. Although this represented a significant increase, the district accepted the fact that the original schedule and effort had probably been underestimated and/or overly optimistic; they also realized that completing implementation of the replacement system was both necessary and beneficial.

4.4 Results

The situation at the district has steadily improved since the successful go-live on both the financial and customer service systems. Although there are still ongoing efforts to retrain and educate staff on the new system and business processes, the district has begun to see the value and benefits of the new financial and customer service systems for the organization. Several of the lessons learned by the district during the ERP replacement project include getting staff up-to-speed and trained on systems as early as possible; don't underestimate the amount of effort that will be spent on data conversion; anticipate many iterations of design, configuration, testing, and rework; rely on your vendors for software expertise, but not on your business processes; and don't rush the system implementation (i.e., take extra time if you need it).

5.0 CASE STUDY 5: MODELING A UTILITY FRAMEWORK

In reference to Chapter 4, Figure 4.5, further elaboration of the conceptual modeling of a water utility is provided in this Case Study 5. In the following illustrative example of Figure 10.11, and referring back to a program's prime interrogatories (i.e., who, what, when, where, why, and how), the "whys" are shown as "motivations," and are directly related to the water agencies' business plan objectives. The "whos" are shown as the "people" who meet the why's objectives. The "hows" are shown as the "function" or business process by which the objectives are met, and the "whats" are shown as the "applications" and "data," which are the tools used to meet the same objectives. Specific labels are omitted here for clarity, although each of these points does, in fact, correlate to actual utility business framework nodes (i.e., MWD). The diagram is shown in Google Earth for the three-dimensional rendering ability.

While these nodes alone are substantial, they do not yet illustrate the connectivity between who does what, how, and with what tools (the tools being the IT systems). However, by looking at only a specific objective of maintaining water quality, as in Figure 10.12, one can begin to discern the linkage of who does it, by which function, and by using what data and applications. This can also be, and has been, linked to actual field assets on the ground. The reality seen here is that Water Quality alone. Yet, even this is part of a broader architecture; water quality is connected to other aspects of the framework; for example, it shares information with accounting on chemical quantities and with environmental regarding regulatory compliance.

As was shown in Figure 4.5, all of the utility's actual business objectives, personnel groups, business process classes, applications, and data sources (i.e., why, who, how, and what) can be extremely complicated, even at a conceptual level. Although this framework is visually overwhelming, it represents the true and actual (even simplified) linkages between data, applications, business processes, and people. Ideally, this knowledge can be used to streamline both business practices and the associated IT frameworks that support them. By referring back to the matrices that generated this framework, both gaps and redundancies in process, function, application, and data can be identified and mitigated.

This figure represents an IT architectural "framework" in the most literal sense. Further development of this model, currently underway, will link data to the specific assets on the ground that generate the data, thus creating a merger of

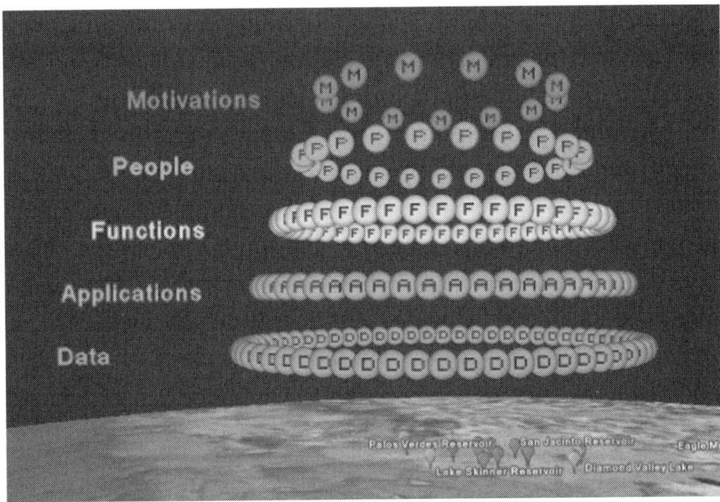

FIGURE 10.11 A utility objective business architecture: prime nodes (courtesy of D. Henry, P.E., MWD of So. CA).

the mutually dependent physical/operational/geospatial and conceptual/business architectures.

A key point to observe in this framework is that the data being collected ARE directly traceable all the way up to objectives or motivations. Conversely, and ideally, motivations drive only those, people, functions, applications, and data needed.

It is also important to note that while the IT layers are really limited to the applications and data, they are the most complicated. Perhaps this means that the IT systems are more complicated than the business itself. In this representation, these are general categories or classes of each group of people, business processes, applications, and data. As these are expanded to their full detail (e.g., actual detailed data vs a general data store), the complexity and interconnectivity can be immense. While this model shows what the agency does every day, it also shows what the IT systems must support every day.

This framework was developed from an internally generated spreadsheet catalog of the aforementioned classes, with linkages established between each class. The graphic was created as a better way to digest and make meaning of a rather large set of data, along with the ability to turn "on and off" or visualize specific objective

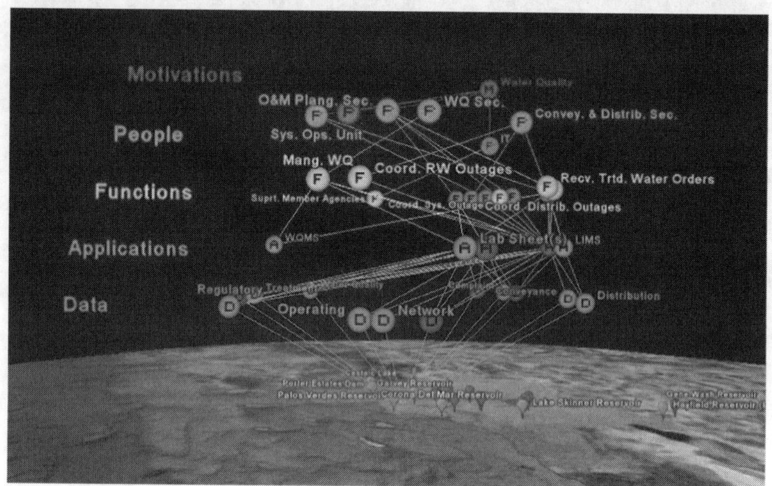

FIGURE 10.12 A utility objective business architecture: water quality (courtesy of D. Henry, P.E. MWD of So. CA).

networks, staff spheres of influence and responsibility, application dependencies, and data propagation (e.g., the water quality-only example in 10.11).

Although the visualization is telling, the data themselves delineated problems and opportunities, namely

- Duplication of data,
- Missing data,
- Data being generated for no apparent reason,
- Duplication of applications,
- Missing process automation,
- Duplication of business processes, and
- Missing business processes.

Although this level of development is not always necessary, the illustration is helpful in showing what it is that IT program managers are really responsible for, that is, the development, installation, integration, and maintainability of data and applications to support business processes that people use to meet business objectives, which can be an extremely complicated undertaking.

6.0 CASE STUDY 6: DEVELOPMENT OF A SPECIFIC BUSINESS PROCESS MODEL, FROM CONCEPTUAL TO DATA MAPPING

In the case of MWD, the process of conceptual business process modeling, followed by data modeling and detailed system specifications, was successfully piloted. The effort resulted in the following benefits:

- This was a cross-organizational process that covered three different business areas. It took several lengthy meetings for these three parties to agree upon the details of the process they had been performing for years. Several gaps and redundancies were identified, and several process objectives and steps were clarified. In the end, a map of the as-is process was agreed upon.
- In the next step, the process was reengineered, streamlined, and improved, and a map of the to-be process was developed. Even if the effort had stopped here, with no software ever developed, the business process, level of knowledge, and coordination were significantly improved.
- From the to-be conceptual diagrams, a set of conceptual data flows and processes were developed (i.e., what data goes in, what is done with it, and what comes out, in business terms). This provides even more understanding of the actual data needs and also provides a useful bridge and communication tool to leverage business user needs and developer instructions. It helps programmers specifically develop what the users need, with much less waste than past processes.
- As a final design step, a set of actual data and system specifications were developed. Although this is more on the technical side, it provides a set of plans from which programmers can work, which significantly boosts efficiency. This also develops rules and data module that can be reused and shared by future applications, cutting future development and maintenance times.
- Finally, prototype software was developed in record time. Besides speed, the benefit is that each rule in the software can be directly traced back to both physical data and conceptual business process and requirements; the structural details support the architectural sketch.

Based on the success of this pilot demonstration, MWD has gone on to purchase professional-grade business process modeling tools that are now used on projects requiring significant process engineering custom-application development projects.

7.0 REFERENCES

Barnett, M.; Lee, T.; Jentgen, L.; Conrad, S.; Kidder, H.; Woolschlager, J.; Cantu, E.; Lozano, S.; Kelly, M.; Eaton, D.; Hollifield, D.; Groff, C. (2004) Real-Time Automation of Water Supply and Distribution for the City of Jacksonville, Florida. *J. Environ. Instrum., Control Autom., Jpn.*, **9** (3), 15–28.

Jentgen, L. A.; Conrad, S.; Kidder, H.; Lee, T.; Barnett, M.; Woolschlager, J. (2005) *Optimizing Operations at JEA's Water System*. American Water Works Association: Denver, Colorado.

Index